Physical Basis of
Plasticity in Solids

Physical Basis of Plasticity in Solids

Jean-Claude Tolédano

Ecole Polytechnique, France

World Scientific

NEW JERSEY · LONDON · SINGAPORE · BEIJING · SHANGHAI · HONG KONG · TAIPEI · CHENNAI

Published by

World Scientific Publishing Co. Pte. Ltd.

5 Toh Tuck Link, Singapore 596224

USA office: 27 Warren Street, Suite 401-402, Hackensack, NJ 07601

UK office: 57 Shelton Street, Covent Garden, London WC2H 9HE

British Library Cataloguing-in-Publication Data

A catalogue record for this book is available from the British Library.

PHYSICAL BASIS OF PLASTICITY IN SOLIDS

ISBN-13 978-981-4374-05-7
ISBN-10 981-4374-05-9

Printed in Singapore.

Preface

This book is based on a physics course first launched in 2002 for the students at Ecole Polytechnique, a French scientific institution accessible after two years of university-level studies. The course was attended by students majoring in a second year program (corresponding to a fourth year in university) entitled "Structure Mechanics", and intending to specialize, after their graduation from Ecole Polytechnique, in various fields such as civil engineering, aeronautics, mechanical engineering, nuclear engineering, or research in the field of materials science. The program included a set of courses on the mechanical properties of solid state materials, among which was an introduction to the methods used in the mechanical sciences to analyze the *plastic deformation* of solids or their *fracture*.

The physics course introduces the physical mechanism of the plastic deformation, that relies essentially on the occurence and motion of *dislocations*. Three of the eight chapters of this book are devoted to the study of these *linear defects* of crystalline solids. Two other chapters provide the elements of crystallography and mechanics required to discuss their properties. One chapter describes other types of crystal defects with which the dislocations interact. The main chapter follows, devoted to the description of the mechanism of plastic deformation, based on the preceding elements.

The appendices contain over thirty exercises and their solutions, which were used during classes or written examinations. Many of these exercises are adapted from texts found in various standard textbooks, as quoted in Chapter 1.

On a topic involving many different aspects and a great variety of materials behaviour, we had to elaborate to eight lectures, each of duration of an hour and a half, and as many two-hour exercise classes. One had to therefore restrict to the main ideas, and show their relevance in interpreting phenomena well known to everyone (Why certain metals are harder than others? Why heating metals makes them more compliant, etc...). It was also necessary to describe the methods used in calculating the properties of dislocations, and train students to practice these calculations.

It appeared that a good understanding of the subject, by the students, could be obtained by the following organization: Chapters 1 and 2 (one course), Chapters 3 and 4 (one course), Chapter 5 (two courses), Chapter 6 (one course), Chapter 7 (one course), Chapter 8 (two courses).

It is worth mentioning that the initial inspiration of this course is rooted in earlier ones on the same matter given to the students of Ecole Polytechnique. Yves Quéré, had first taught the subject in 1989. His course was based on two chapters of his book "Physique des Matériaux" (Ellipses editor 1988, English translation "Physics of Materials" Gordon and Breach 1998). The teaching of the subject was continued by Denis Gratias, between 1994 and 2002, relying on his lecture notes "Introduction à la Physique des Matériaux" (unpublished). These two texts cover a larger set of topics than the present book. In particular, they devote substantial attention to the metallurgical phase changes and to the mechanism of fracture.

At the end of Chapter 1, we have listed a few reference books, written in French or in English, and discussing more thoroughly the subjects considered in this book. Most of these references were written during the years 1960−1970 (with, sometimes, revised versions published later), that is, towards the end of the period of clarification of the central ideas relative to dislocations and to their role in the mechanism of plasticity. With the exception of the above mentioned works, by Yves Quéré and Denis Gratias, and of the treatise by D. Hull indicated at the end of Chapter 1, most listed books are rather aimed at researchers than at undergraduate students.

Collaborating with Denis Gratias during his teaching of this subject between 2000 and 2002 was invaluable for me to outline the composition of this course. I am indebted, to several colleagues with whom I have taught, for their useful remarks, namely Alain Barbu, François Beuneu, Yann Le Bouar, and Jean-Eric Wegrowe. I am grateful to my colleagues from the Department of Mechanics, Dominique Barthes-Biesel, Pierre Suquet, and André Zaoui for their friendly welcome to their "structure mechanics" program. Finally, I would like to thank Jean-Paul Coard for encouraging me to write the French edition of this course.

Jean-Claude Tolédano
Paris, April 2011

Contents

Chapter 1

Introduction

1.1 Plasticity

Metals, and, to a lesser extent, most solid materials, can undergo a *permanent* change of shape when submitted *temporarily* to external forces of sufficient magnitude. This mechanical property is called *plasticity*. It has been used since the beginning of the Bronze Age, in order to manufacture tools or weapons by turning pieces of metals into desired shapes. This was achieved with the help of bending or hammering forces, this action being made more efficient through heating the material to high temperatures. The present book aims at presenting the main ideas which constitute the *microscopic* physical explanation of the plastic behaviour of solids.

The explanation will refer, almost exclusively, to the case of *crystalline solids*, i.e. solids which are built from a spatially periodic assembly of atoms. Two reasons justify this restriction.

The main one is that the atomic-scale periodicity of crystals is an essential element of the theory which accounts for the plastic behaviour of solids. Indeed, this theory, developed progressively between 1920 and 1960, gives a central role to specific defects of crystals. These defects, which are called *dislocations*, are put in motion when suitable *external forces* are applied to a solid. This motion as well as the interaction of dislocations with different types of *internal forces* due to other defects, or to the atoms constituting the solid, are the basic ingredients used to analyze the various characteristics of the plastic behaviour.

Another reason is that *metals and alloys*, which are the solids displaying in the most spectacular way the plastic behaviour, are crystals. More precisely, a piece of a metal or alloy is, generally, an assembly of *grains*, each grain being a crystal of micronic size, whose individual properties determine, to a large extent, the mechanical properties of the metal or the alloy.

In order to further clarify the object of the book let us recall some characteristics of the mechanical properties of solids.

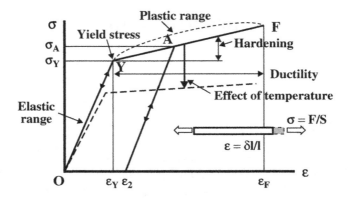

Figure 1.1   Schematic relationship between stress and strain in a solid material. OY is the elastic range and YF the plastic range.

1.1.1   *Mechanical properties of solids*

Figure 1.1 summarizes the schematic mechanical behaviour of a solid material. It shows the complex relationship between the mechanical stress σ applied to a solid and the strain (also called deformation) ϵ resulting from this application. In a simplified approach, this relationship is deduced, for instance, from the measurement of the relative change of length $\epsilon = \delta l/l$ of a rod of section S submitted to a force $F = \sigma.S$ pulling on the ends of the rod. Five aspects of this relationship can be emphasized.

i) When the magnitude σ of the stress is smaller than a value σ_Y , called the *yield strength* or alternately, *the elastic limit,* the mechanical behaviour of the solid is *reversible.* Thus, the value of the strain $\epsilon(\sigma)$ is the same for increasing values of the stress or for decreasing ones. In particular, if the value of the stress is brought back to zero, the induced strain vanishes. This behaviour corresponds to the *elastic* range labelled OY of the curve plotted on Fig. 1.1. In this range, one can generally consider, to a good approximation, that the strain is proportional to the stress: $\sigma \simeq C\epsilon$.

The C coefficient of proportionality between the strain and the stress, is a measure of the *elastic stiffness* of the considered solid material. It is an important characteristics of its mechanical properties. For metals, the value of the stiffness lies in the range 10^{10}-10^{11}Pascals ($\sim$1-10 tons per square millimeter). Such a value means that a weight of one kilogram suspended to a wire of section one square millimeter will determine a relative elongation $\delta l/l$ of the wire of $10^{-5} - 10^{-6}$.

As for the the yield strength σ_Y, it characterizes the *hardness* of a solid material. Its value does not only depend of the nature of the material, but also, in a pronounced manner, of the mechanical and thermal processing imposed to it, as well as of other parameters such as, for instance its temperature. In a "soft" material, σ_Y can be less than 10^6 Pascals ($100g/mm^2$), whereas in a "hard" one it can be higher than 10^9 Pascals ($100 kg/mm^2$). A current value of the strain ϵ_Y corresponding to the yield strength is $\sim$1%.

ii) For $\sigma > \sigma_Y$, the material becomes *plastic*. In this range of σ values (YF on the plot) the deformation becomes non-reversible. Bringing the value of the stress down to zero does not cancel the strain. For instance, starting from point A, the strain decreases along the line $\mathbf{A}\epsilon_2$, and a permanent strain ϵ_2 remains at zero applied stress.

iii) The application of a moderate stress to the material initially in the state ($\epsilon = \epsilon_2, \sigma = 0$), determines, again, a reversible evolution along $\epsilon_2 A$ with a quasi-linear relationship $\sigma \simeq C'(\epsilon - \epsilon_2)$ [1] With a further increase of the stress, a new elastic limit σ_A is reached. The schematic situation represented on Fig. 1.1, in which YA has a positive slope, characterizes a material in which the elastic limit σ_A after deformation is larger than σ_Y. The material is *hardened* by the plastic deformation.

iv) The end point F, corresponds to the *fracture* (or rupture) of the solid-sample, i.e. the loss of its cohesion. It is reached after a permanent elongation ϵ_F. In certain metals, one can elongate a rod to one hundred times its initial length before fracture occurs. Such a metal is very *ductile*. The ductility of a material is associated to the value of the deformation ϵ_F. The larger ϵ_F, the larger the ductility. When almost no ductility exists the material is *brittle*. In this case, the fracture occurs just above the yield strength (ϵ_Y, σ_Y) and F then almost coincides with Y. Similarly to the yield strength, ductility is a function of the "mechanical history" of the sample considered, as well as of its temperature.

---

[1]On the plot, the rigidity $C\prime$ has been represented as equal to C: the line $\epsilon_2 A$ is parallel to OY. This corresponds to the simplest type of behaviour in which the rigidity is approximately the same in all the reversible ranges. This behaviour will be justified in chapter 8 by the fact that the perfection of the crystalline order is almost preserved in the volume of the material.

v) Temperature modifies significantly all the mechanical properties of a material. In general, heating a solid makes easier its plastic deformation: it decreases the value of the yield strength (ϵ_Y, σ_Y), and it increases the ductility. It also decreases the hardening (the slope of YA).

The table below provides a few examples of values measured in currently used solid materials at room temperature.

Material	longitudinal stiffness (10^{10}Pa)	yield strength (10^7 Pa)	
Lead	5	$\sim$0,01	
Aluminum	11	$\sim$0,1	
Copper	17	$\sim$0,1	(1.1)
Gold	19	$\sim$0,1	
Iron	24	1-10	
Silicon	17	10-100	
Diamond	100	>100	

1.1.2 *Microscopic mechanisms*

In the above macroscopic description (cf. Fig. 1.1), the elastic and plastic deformations of a solid appear as two successive steps of a continuous process. Actually, these sequences have a very different microscopic background. To grasp their difference of nature, it is worth giving, prior to a description of the plastic behaviour, some indications on the microscopic origin of the elastic behaviour.

Elastic behaviour

The interpretation of the elastic behaviour of a solid, from the standpoint of its atomic structure, derives from a simple principle. Thus, the assembly of atoms constituting a solid possesses an *equilibrium spatial configuration*, in which the interatomic distances are well defined.[2] In the absence of external forces applied to the solid, any deviation of atomic configuration (Fig. 1.2), with respect to the equilibrium one, will increase the total energy of the solid, and has therefore a tendency to regress. Such a deviation, consisting in small changes in the distances between atoms, will actually

---

[2]At very low temperatures this configuration corresponds to the minimum of the energy of the set of particles composing the solid.

be induced by the macroscopic deformation, imposed to the solid by an external stress. When the stress is suppressed, the atomic configuration will tend to return to equilibrium, thus accounting for the reversibility of the behaviour. Moreover, as the deviations provoked by strains are very small (less than a percent) the effective forces acting on the atoms to bring their distances back to their equilibrium values, will therefore be appproximately proportional to the small deviations, thus justifying the linearity of the elastic behaviour.

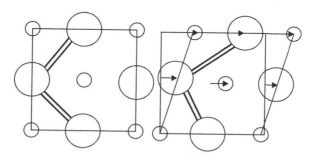

Figure 1.2   Schematic representation of the atomic displacements induced by a macroscopic shear. There are changes of the interatomic distances and of the angles between chemical bonds. Restoring forces are generated which tend to bring back the distances and angles to their initial equilibrium values.

The simplicity of this explanation contrasts with the difficulty of its effective working out. To calculate the elastic stiffness of a specific material, whose atomic configuration is known, is very complex. Indeed, mechanical deformations are associated to *collective displacements* of a large number of atoms interacting with each other. The calculation, which has to use quantum mechanics, has to take into account this large number of interactions. This situation explains the fact that such a calculation could only be achieved in recent years (using computerized procedures) for materials with the simplest atomic configuration.

Plastic behaviour

The interpretation of the plastic behaviour involves several ideas, all of them important.

First, there is a suggestion made in the beginning of the years 1920 (**Polanyi and Schmid**) which stated that the permanent elongation of a rod induced by an axial traction, is, in fact, the cumulative result of a series of small glidings (also called slips). Each portion of the rod glides with respect to the adjacent portion. All the slips are parallel to plane directions which are inclined with respect to the axis of the rod. (Fig. 1.3).

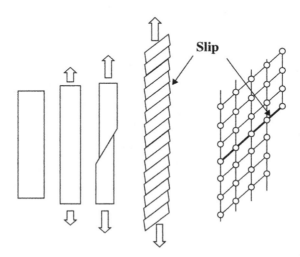

Figure 1.3   Successive steps of the vertical plastic elongation of a rod submitted to a traction parallel to its axis. (Left to Right) 1) Undeformed rod, 2) Elastic elongation, 3) Beginning of the plastic deformation by relative glides (or "slips") of the upper part of the rod along a plane inclined with respect to the axis, 4) multiplication of the glides along parallel planes. The addition of the glide-projections on the vertical axis determine the observed elongation of the rod. On the right part of the figure, the elementary glide (slip) at the atomic scale, is represented. Its amplitude is equal to one interatomic spacing, and its direction is parallel to a lattice plane.

The *directions of the planes* are those of specific atomic planes of the crystalline material considered. Moreover, the *amplitudes* of the small slips, less than one micron, are exact multiples of the *crystal periods* (of Angström order of magnitude) characteristic of the spatial periodicity of the crystalline material.

At the atomic scale, the plastic deformation thus appears as a *discontinuous phenomenon*, consisting in a succession of elementary slips having an amplitude equal to one crystal period in a definite direction, whose characteristics are hence closely related to the spatial configuration of the atoms in the material considered.

A second idea relies on the observation that the measured values of the yield strength σ_Y (in the range $10^6 - 10^8$ Pascals) are much smaller, by at least an order of magnitude, than the forces which bind the atoms together within the material. This observation has led in the years 1930-1940, to the assumption that plasticity is only possible because the crystalline nature of the material is *imperfect*. As already mentioned above, it contains *linear defects*.[3] These defects make easier the relative gliding of adjacent portions of the material, and thus reduce the forces necessary to permit the gliding. In the framework of this theory, the elementary step of a plastic deformation, which consists in the slip, by one atomic period, of one portion of the material, corresponds to the *motion* through the entire width of a rod, *of one such linear defect*

Interestingly enough, the existence of the linear defects and their role in the plastic deformation, have been suggested theoretically already in 1934 (**Orowan, Polanyi** and **Taylor**), while these ideas were experimentally confirmed (Fig. 1.4) only after 1956 (**Bollmann** and **Hirsch**). Conversely, it could be checked that in crystals specially prepared to be *free from such defects* the yield strength is, conclusively, considerably larger than the values currently observed.

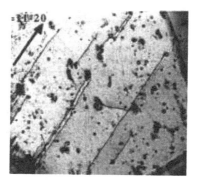

Figure 1.4   Observation by electron microscopy of quasi-straight lines of dislocations, a few microns in length, in zirconium (photographs obtained by Franck Ferrer; Thesis 2000 Ecole Polytechnique).

---

[3]Linear defects are quasi-infinite (i.e. very large with respect to the atomic scale) in one dimension, and confined to a few atomic sizes in the two other dimensions (Cf. also chapter 4)

It was also recognized that these linear defects had the same properties as *singular lines* defined, in continuous media, by mathematicians, in the beginning of the twentieth century (**Weingarten 1901, Volterra 1907**), which had later been called *dislocations* (**Love 1920**).

Other important conceptual developments pertain to the relationship between, on the one hand, the different characteristics of the plastic behaviour (value of the yield strength, magnitude of the plastic deformation, hardening etc...), and, on the other hand, the interactions between the dislocations and various types of objects.

Thus, a simplified "continuous" theory of dislocations could account for the interactions between a dislocation and the internal stress in a material (**Burgers 1939, Peach** and **Koehler 1950**), between several dislocations, between dislocations and point defects (vacancies, impurities, clusters, etc...), between dislocations and planar defects (such as grain boundaries). These various interactions govern, in particular, the mechanisms of generation of dislocations (**Frank** and **Read 1950**), which are central features of the theory of plasticity.

Finally, models of the atomic configuration of a dislocation, and of its interaction with the crystal atomic potential (**Peierls, Nabarro 1940**), have enabled an estimation of the value of the stress needed to put in motion a dislocation.

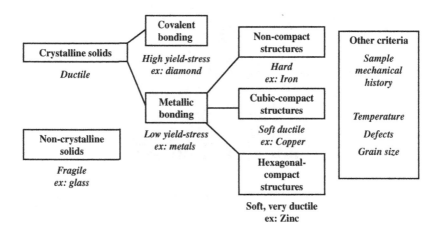

Figure 1.5   Main elements which determine the diversity of plastic properties of different solid materials. The atomic structure is a central feature.

1.2   Organization and contents of the chapters

The scope of this treatise is, mainly, to present an analysis of the phenomena occuring at the microscopic scale (the atomic scale, but also a larger "mesoscopic" scale corresponding to dislocations) which underly the general characteristics of the plastic behaviour, namely their hardness, their ductility, and the effects of temperature.

Attention will also be given to the elements which allow to understand the diversity of solid substances with respect to their mechanical properties. Namely the large difference in hardness between different metals, their relative softness as compared to non-metallic solids such as diamond, or the fact that the effect of temperature is much more pronounced in certain materials.

Figure 1.5, anticipates on the following chapters by showing some of the elements which determine the specific mechanical properties of a given material.

It is remarkable that the relative positions of atoms in space (the so-called "atomic structure") has a central role. This justifies that the book includes, prior to any physical consideration, an introduction to the geometry of crystals at the atomic scale. Hence, the next chapter introduces the few elements of crystal geometry which are needed to describe dislocations and other types of defects relevant to the study of plasticity. Namely, *crystal lattices and translations, lattice planes, unit cells, and symmetries.* On the other hand, a few simple atomic configurations, frequently encountered in metals, are described. Also, indications are given of the structure of non-crystalline solids which will later clarify the mechanical differences between these materials and metals.

Chapter 3 is a recall of the basic notions of the mechanics of continuous media required in the study of plasticity.

Chapter 4 is devoted to the configuration of the so-called "real crystalline solids" which always involve *defects* of the periodicity. The various types of defects are enumerated. A specific attention is then given to a simple *point defect*, the vacancy, which consists in the absence of an atom. This type of defect has an important role to explain the plastic behaviour at high temperatures.

Chapter 5 concerns the study of *dislocations.* Two different approaches are used. The first one focuses on its nature of linear defect of the atomic configuration. The other considers the dislocation as a line of singularities in an elastic continuous medium. In both cases, one can characterize their geometrical and physical properties by a so-called *Burgers vector*, which is a vector of defined direction and modulus. A classification of dislocations will be introduced, based on the angle made by the Burgers vector with the dislocation line. The simple cases of the *edge* and *screw* dislocations will be considered as well as the case of a *dislocation loop*. The principles of observing dislocations by means of the use of electron microscopy will also be given attention.

In chapter 6 dislocations are considered as objects of an elastic continuous medium. In this framework, the strain and stress fields generated by a dislocation are studied. Conversely, the action of a stress field on a dislocation is determined. The latter result is the key to the study of the interaction between dislocations or between a dislocation and a point defect.

Chapter 7 describes a model of the microscopic structure of a dislocation, with the view of studying its interaction with the crystal atomic potential. This interaction underlies the determination of the forces needed to put in motion a dislocation.

Chapter 8 deals with two subjects. In the first place, the mechanisms of generation of dislocations are considered, as well as the conditions of their mobility. One is then able, on the basis of the preceding chapters, to analyze the general principles governing the plastic behaviour of a solid material, as well as the dependence of this behaviour on the chemical and structural nature of a solid and of its temperature.

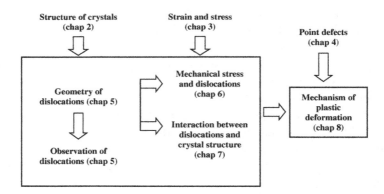

Figure 1.6   Organization of the chapters leading to the physical explanation of plastic properties. The large frame contains the chapters relative to the properties of dislocations.

Figure 1.6 summarizes schematically the organization and contents of the different chapters.

1.3 General References

A.H. Cottrell, *The Mechanical Properties of Matter* (John Wiley, New-York 1964).

J. Friedel, *Dislocations.* (Pergamon Press, Oxford 1964).

F.R.N Nabarro, Z.S. Basinski, D.B. Holt *The Plasticity of Pure Single Crystals* (Advances in Physics vol. 13 N°50, London 1964).

D. Hull, *Introduction to Dislocations* (Pergamon Press, Oxford 1965, revised 1981)

J.P. Hirth, J. Lothe, *Theory of Dislocations.* Second Edition (Krieger Publishing, Malabar Florida 1982)

F.R.N. Nabarro, *Theory of Crystal Dislocations* (Dover, New-York 1987).

Y. Quéré, *Défauts ponctuels dans les métaux.* (Masson editor, Paris 1967).

W.T. Read, *Les dislocations dans les cristaux* (Dunod, Paris 1957)

E. Braun, *Mechanical Properties of Solids* (Article pertaining to the formation of ideas in this field) in *Out of the Crystal Maze,* Editors L. Hoddeson *et al.* (Oxford University Press 1992).

Y. Adda, J.M. Dupouy, J. Philibert, Y. Quéré *Eléments de métallurgie physique* **Vol 3 and Vol 5** (Edit. INSTN-CEA, 2000).

J-L. Martin, J. Wagner, *Dislocations et plasticité des cristaux* (Presses Polytechniques et Universitaires Romandes, Lausanne 2000).

Y. Quéré, *Physique des Matériaux* (Editions Ellipses, Paris 1988).English edition, *Physics of Materials* (Gordon and Breach, London 1998).

D. Gratias *Introduction à la physique des matériaux* (Course of Ecole Polytechnique, Palaiseau 2001).

Chapter 2

The structure of crystalline solids

Main ideas: Algorithm of construction of a crystal (lattice + basis). Geometrical implications of the three-dimensional periodicity (lattice planes and rows), restrictions on the values of rotation angles and on the unequivalent types of Bravais lattices. Different types of unit cells. Simple packings of atoms.

2.1 Introduction

Considered at the atomic scale, solids are part, as well as liquids, of the condensed matter systems, in which the atoms are in "contact" with eachother. Their well defined external shape, in given conditions of stresses and of temperature, is related to the fact that the equilibrium position of each atom, referred to a frame attached to the solid, is "fixed".[1]

From the standpoint of the spatial configuration of their constituting atoms, several types of solids exist. One distinguishes *crystalline* solids from non-crystalline ones (amorphous, quasi-crystalline, etc...). For reasons stated in the introductory chapter, it is mainly the crystalline solids which will be described in this chapter. In these systems, observations by means of instruments giving access to the atomic-scale, show the occurence of *specific geometrical regularities* in the relative positions of the microscopic constituents.[2] Understanding these regular patterns is a necessary step

---

[1]The correctness of the preceding statement requires mentioning that the concerned atomic positions are average positions of the atomic nuclei. Indeed, the atoms in a crystal are in constant motion, vibrating about a point which is their average equilibrium position, with amplitudes of the order of 5.10^{-2}Å at room temperature. The measurements of these positions (by X-ray or microscopic techniques) are performed on time scales which are very large with respect to the periods of the vibrations. They therefore reveal, consistently with the above definition, the average atomic positions.

[2]The main scientists who have analyzed these regularities are the French Haüy and Bravais, the Germans Hermann and Schoenflies, and the Russian Fedorov.

in the study of the mechanical properties of solids. The present chapter is devoted to the description of these regularities which are also termed *symmetries*.

2.2   Crystal geometry

2.2.1   *Ideal crystal*

The *structure* of a solid is defined as the configuration in space of its constituting atoms.

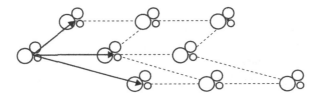

Figure 2.1   Starting with a "basis" of atoms (here a group of three atoms), a crystal is constructed by placing an identical basis at the end of each vector of the form $(n_1 \vec{a}_1 + n_2 \vec{a}_2 + n_3 \vec{a}_3)$.

This configuration can be specified by the set of coordinates of the nuclei of all the atoms. It is useful, in a first approach of description of "real solids", to define the concept of *ideal crystal*. Such a crystal is an infinitely extended solid, the structure of which possesses a *three-dimensional spatial periodicity*. By this statement, one implies that the structure can be built by *replicating* a given set of atoms, through displacements defined by *translations* $\vec{T}$ of the form:

$$\vec{T} = n_1 \vec{a_1} + n_2 \vec{a_2} + n_3 \vec{a_3} \tag{2.1}$$

where the coefficients n_1, n_2, n_3 are integers (positive, negative, or equal to zero), and where the three vectors $\vec{a_1}, \vec{a_2}, \vec{a_3}$ are *three non-coplanar* vectors. These vectors define the *periodicity* of the crystal. They are its *primitive* (or elementary) *translations*. The set of atoms, the replication of which generates the crystal, is the *basis* of the structure. For many crystals of common use, such as metals, the basis contains a single atom or a few atoms. Owing to the fact that the atoms of a crystal are in contact with eachother, the lengths of the primitive translations and the linear size of the basis are in the same range of values. Thus, when the basis is reduced to a single atom (as is the case in simple metals) the moduli $|\vec{a}_i|$ are equal to ~2,5-3Å, i.e. the diameter of an atom.

The manner of generating a given crystal structure from a basis and a set of translations is not unique. One can replicate a given basis, or alternately, replicate a larger group of atoms constituted by a set of several such bases. In the latter case, certain of the three elementary translations will be multiples of the primitive translations considered initially. The ambiguity of the given definition is lifted by specifying that the basis is the smallest set of atoms allowing such a construction.

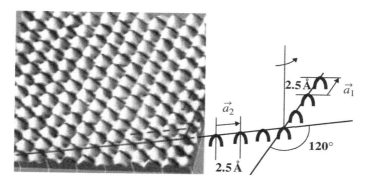

Figure 2.2   Atomic-scale image of the surface of a graphite crystal obtained with an "atomic force" microscope (Courtesy of J.M. Moison 1990). Each "mound" is a group of two carbon atoms. The schematic drawing at the right emphasizes the symmetry of the structure with respect to 120° rotations.

It is worth pointing out that the translations $\vec{T}$ in Eq. (2.1) have been defined as a *tool of construction* of the structure through replication of the basis. A different point of view can be adopted. The structure being infinitely extended, its *overall displacement* by any of the translations (2.1) will consist in bringing a given set of atoms to a position where it substitutes an identical set. The structure is then *transformed* into one which is superimposed on the initial one and is undistinguishable from it.

From this standpoint, each translation $\vec{T}$ in Eq. (2.1), is viewed as a geometrical transformation acting in the three-dimensional space, and leaving the atomic structure of the crystal globally *invariant* (unchanged). $\vec{T}$ defines the *translational symmetry* of the structure.

Such a point of view can be extended to other geometrical transformations than translations. For instance, examination of the two-dimensional structure represented on Fig. 2.2 shows that a rotation by an angle of 120 degrees about an axis passing through the origin, and perpendicular to the surface of the crystal, exchanges identical constituents and preserves globally the structure.

The preceding remark has a general validity. In addition to the translations in Eq. (2.1), certain specific rotations or other transformations (e.g. mirror symmetries) can leave globally unchanged the atomic configuration of a crystal. *The complete set of these symmetry transformations* (in which the translations (2.1) are included), defines the so-called *space-symmetry* of the crystal. It is this set of symmetry transformations which substantiates the idea of an atomic-scale regularity in crystals.

2.3   Bravais lattices

2.3.1   *Definition*

The Bravais lattice (Fig. 2.3) associated to a given crystal structure is a three dimensional, infinite, set of points formed by the end-points $M(n_1, n_2, n_3)$ of all the vectors :

$$\overrightarrow{OM}(n_1, n_2, n_3) = \overrightarrow{T}(n_1, n_2, n_3) = n_1\overrightarrow{a_1} + n_2\overrightarrow{a_2} + n_3\overrightarrow{a_3} \qquad (2.2)$$

where O is an origin. Each point M of this lattice is called a *node*.

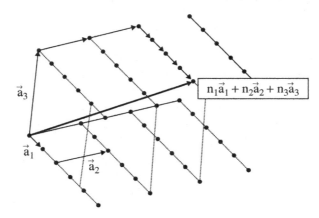

Figure 2.3   Primitive translations of a crystal and Bravais lattice formed by the set of end-points of the crystal translations.

Two different meanings can be attached to the Bravais lattice of a crystal. First, this lattice can be considered as *part of the structure,* by chosing its origin O within the structural basis, e.g. on the nucleus of a specific atom of the basis. All the nodes will then be located on the nuclei of identical atoms. From this standpoint, the Bravais lattice constitutes the underlying skeleton of the structure. Putting "flesh" on this skeleton in the view of obtaining the actual structure consists in placing at each node M a set of atoms related to the basis by the displacement Eq. (2.2).

On the other hand, the Bravais lattice can be considered as an abstract entity, a mere set of points of arbitrary origin, separate from the structure of the crystal, but *associated* to it, and providing a *geometrical picture of the set of translations* $\overrightarrow{T}$ defining the periodicity of the crystal. This standpoint is convenient for deriving the rules governing the symmetry properties of crystals.

2.3.2 *Properties*

Non-unicity of the generating translations

Figure 2.4 provides an example of a two-dimensional lattice of nodes obtained by means of a formula of the type Eq. (2.2), involving two vectors $\overrightarrow{a_1}$ and $\overrightarrow{a_2}$. Obviously, all the nodes of this lattice can also be generated by using the two basic vectors $\overrightarrow{a'_1} = (2\overrightarrow{a_1} + \overrightarrow{a_2})$ and $\overrightarrow{a'_2} = (\overrightarrow{a_1} + \overrightarrow{a_2})$ or, alternately, the two basic vectors $\overrightarrow{a"_1} = \overrightarrow{a_1}$ and $\overrightarrow{a"_2} = (-\overrightarrow{a_1} + \overrightarrow{a_2})$.

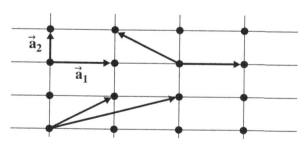

Figure 2.4   Generation of the same Bravais lattice by different sets of primitive translations.

Likewise, it is easy to check that the Bravais lattice of any three dimensional crystal can be generated either by the set of vectors $(\overrightarrow{a_1}, \overrightarrow{a_2}, \overrightarrow{a_3})$ or, for instance, by the set $(\overrightarrow{a_1} - \overrightarrow{a_2}, \overrightarrow{a_2}, \overrightarrow{a_3})$. Hence, several choices of primitive translations are possible to generate the uniquely defined lattice of points underlying the structure of a crystal.

Lattice planes and rows

A plane defined by any three, non-aligned, nodes is a *lattice plane*. It can be shown that such a plane contains an infinite number of nodes which form a two-dimensional Bravais lattice, i.e. the two-dimensional set of end-points of the vectors:

$$\overrightarrow{OM}(n_1, n_2) = \overrightarrow{T}(n_1, n_2) = n_1 \overrightarrow{c_1} + n_2 \overrightarrow{c_2} \tag{2.3}$$

$\overrightarrow{c_1}$ and $\overrightarrow{c_2}$ being combinations of the $\overrightarrow{a_i}$ with integer coefficients. All the nodes of a Bravais lattice can be grouped on a set of equally spaced parallel planes constituting a family of lattice planes. This distribution of the nodes into a stack of parallel planes can be achieved, in an infinity of manners, by stacks having different orientations. Each family, defined by the orientation of the perpendicular vector $\overrightarrow{u}$ corresponds to a specific spacing $d(\overrightarrow{u})$ between planes. Lattice planes with different orientations do not *generally* contain the same planar density of nodes (unless they are symmetry related): the most dense planes are obtained by selecting three nearest-neighbour nodes in the lattice. The denser a lattice plane, the larger the associated spacing between planes, since the decomposition of the three-dimensional lattice into different families of lattice planes having different planar densities must determine the same number of nodes in a given volume.

Similarly, all the nodes of a Bravais lattice can be grouped, in an infinity of manners, on a set of equally spaced parallel lines constituting a family of *lattice rows*. A discussion similar to the preceding one would hold for the definition of lattice rows, the direction of which is determined by the selection of any two nodes of the lattice.

Symmetry of the Bravais lattice

The geometrical transformations (Fig. 2.6) which define the space-symmetries of crystals are translations, rotations, mirror symmetries (also called reflexions), the symmetry about a point (also called inversion), as well as combinations of the former transformations. Two of these combinations are worth specifying, namely the *screw rotation* which consists of a rotation combined with a translation parallel to the rotation axis, and the

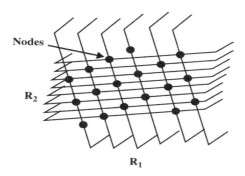

Figure 2.5 Decomposition of the Bravais lattice into a stacking of two-dimensional lattices of nodes on equally spaced planes. R_1 and R_2 are two examples. A third example is the family of planes parallel to the figure.

glide-plane which consists of a symmetry about a plane combined with a translation parallel to the plane.

It can be shown that any of the preceding geometrical transformation S can be denoted

$$S = \{R_O | \vec{t}\} \tag{2.4}$$

R_O is a *point-symmetry* transformation i.e. a pure rotation, a reflexion, an inversion, or a combination of these transformations. R_O leaves unmoved the origin O of the lattice (e.g. O is a point common to the rotation axes). On the other hand, $\vec{t}$ is a translation applied *after* R_O.

In the framework of this notation, the lattice-translations $\vec{T}$ are expressed as $\{0|\vec{T}\}$ (rotation of angle $\theta = 0$ followed by the translation $\vec{T}$).

An important property consists in the fact that if O is a node of the Bravais lattice, this lattice is transformed into itself by the set of all the *point-symmetry* transformations R_O.[3] For instance, if the structure of a crystal is unchanged by a transformation consisting in the succession of a $\pi/2$ rotation around the z-axis, *and* of a definite translation $\vec{t}$, the Bravais lattice is necessarily invariant by the *sole rotation*, applied about any of its nodes, and is therefore "square-shaped".

This property derives from the fact that if a crystal-structure is unchanged by $\{S_O|\vec{t}\}$, its Bravais lattice referred to the same origin O will also be unchanged. Applying the $\{S_O|\vec{t}\}$ transformation to the Bravais

---

[3]In other terms, the set of translations defining the periodicity of the crystal, must be *compatible* with the other symmetry transformations of the structure.

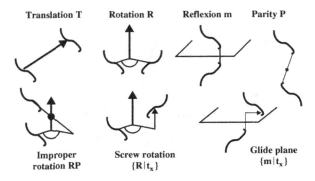

Figure 2.6   Symmetry transformations of a crystal structure.

lattice consists in modifying the directions of the vectors generating the lattice by the point transformation S_O, and, then, in shifting the origin of the lattice by application of the translation $\overrightarrow{t}$. Since the origin of the Bravais lattice is arbitrary, we can cancel this shift and thus bring back the origin to its initial position. The initial and final set of points both coincide with the uniquely defined Bravais lattice of the invariant structure. This lattice is therefore unchanged by the transformation S_O.[4]

One implication of this property of Bravais lattices is the constraint imposed to the angle θ of a *rotation R_O*.

Constraints on the rotation angles

Consider first, in the three-dimensional space, a frame of reference adapted to the Bravais lattice, and consisting of the set of primitive translations $\overrightarrow{a_1}, \overrightarrow{a_2}, \overrightarrow{a_3}$ (Fig. 2.7).

Let $R_O(\overrightarrow{u}, \theta)$ be a rotation of angle θ around an axis $\overrightarrow{u}$ passing through the origin. Since we assume that R_O leaves the Bravais lattice globally invariant, the end-points M_1, M_2, M_3 of the $\overrightarrow{a_i}$, which are nodes of the lattice, are transformed by the rotation into nodes M'_1, M'_2, M'_3 of this lattice. The latter statement can be expressed differently: the $\overrightarrow{a_i}$ are transformed into linear combinations of the $\overrightarrow{a_i}$ with *integer* coefficients. Hence, in this reference frame, the matrix $M(R_O)$ which represents the tranformation of the

---

[4]Note that S_O is not a symmetry of the structure (while $\{S_O | \overrightarrow{t}\}$ is such a symmetry) and that, conversely, $\{S_O | \overrightarrow{t}\}$ shifts the origin of the Bravais lattice and is not one of its symmetries.

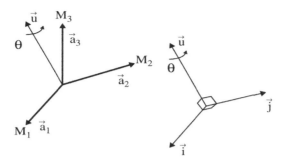

Figure 2.7   Reference frames respectively adapted to the primitive translations and to the axis of rotation.

basic-vectors of the frame is composed of integers. Accordingly, the *trace* of this matrix (the sum of its diagonal elements) is an integer N.

Consider now an *orthogonal* reference frame (Fig. 2.7) defined by the unit vector $\vec{u}$ of the rotation axis, and two vectors $\vec{i}$ and $\vec{j}$ lying in the plane perpendicular to $\vec{u}$. In this frame, the trace of the matrix associated to a rotation R_O is $(1+2\cos\theta)$. The example below is relative to a rotation of $2\pi/6$ around the z-axis in a hexagonal lattice.

$$(\vec{a_1}, \vec{a_2}, \vec{a_3}) \rightarrow \begin{bmatrix} 0 & 1 & 0 \\ -1 & 1 & 0 \\ 0 & 0 & 1 \end{bmatrix} \qquad (\vec{i}, \vec{j}, \vec{u}) \rightarrow \begin{bmatrix} \cos\frac{2\pi}{6} & \sin\frac{2\pi}{6} & 0 \\ -\sin\frac{2\pi}{6} & \cos\frac{2\pi}{6} & 0 \\ 0 & 0 & 1 \end{bmatrix} \qquad (2.5)$$

The value of the trace of a matrix being independent of the reference frame chosen, one can write:

$$(1 + 2\cos\theta) = N \qquad (2.6)$$

Solving this equation leads to a limited set of possible values for $\cos\theta$, namely $(0, \pm\frac{1}{2}, \pm 1)$. Thus, the possible values of the angle of a rotation R_O are:

$$\theta = 0, \pm\frac{\pi}{3}, \pm\frac{\pi}{2}, \pm\frac{2\pi}{3}, \pm\pi \qquad (2.7)$$

Any other value of the rotation angle such as, for instance, $\pi/6$, $\pi/5$, or $\pi/4$ is excluded.

2.4   Unit cells

2.4.1   *Primitive unit cells*

Consider (Fig. 2.8) the parallelepipedic body defined by the three primitive translations $\vec{a_1}, \vec{a_2}, \vec{a_3}$. This body constitutes a *primitive (or elementary) unit cell* of the crystal. Displacing the unit-cell by all the translations $\vec{T}$ of the form (2.1), generates a set of identical adjacent cells, and realizes a *paving* of the three-dimensional space, i.e. a filling, without overlap, of the entire space. Shifting the origin of the paving by $(\vec{a_1}/2, \vec{a_2}/2, \vec{a_3}/2)$ with respect to a node, results in positioning a lattice-node at the center of each cell. Hence, there is a one-to-one correspondance between the nodes and the unit-cells, the density of the nodes in space being equal to one node per volume of the primitive unit cell.

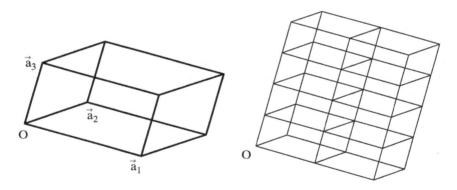

Figure 2.8   Primitive unit-cell of a crystal and paving of space by a set of adjacent cells.

We have pointed out (Fig. 2.4) that the set of primitive translations of a crystal is not uniquely defined. Consequently, there is also an arbitrariness in defining a primitive cell. However, all the possible primitive cells have the same volume, owing to their one-to-one correspondance with the nodes of the Bravais lattice.

2.4.2   *Conventional unit cells*

One can generalize the definition of a unit cell by keeping two properties of primitive unit-cells : a) Parallelepiped based on the lattice translations;

b) Paving of the space by an infinite set of identical bodies; These two properties are preserved in the definition of the *conventional unit-cell*. The conventional cell , unlike the primitive-cell, *is uniquely defined* for a given crystal.

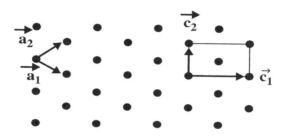

Figure 2.9 Primitive and conventional cells for a rectangular centered Bravais lattice.

In order to understand the advantage of recurring to alternate definitions, let us consider the example of a two dimensional "rectangular centered" Bravais lattice (Fig. 2.9). This lattice is invariant by mirror symmetries through the vertical and horizontal planes passing through any node. The conjunction of these two symmetries is characteristic of a *rectangle*. However, the primitive unit cell is not a rectangle but a diamond, the edges of which are two half diagonals of the rectangle. A rectangular cell can be constructed by using as edges the two lattice translations :

$$\vec{c_1} = (\vec{a_1} + \vec{a_2}) \qquad\qquad \vec{c_2} = (\vec{a_2} - \vec{a_1}) \qquad\qquad (2.8)$$

If one applies to this rectangular cell the translations defined by all the linear combinations of $\vec{c_1}$ and $\vec{c_2}$ with integer coefficients, one generates a set of identical adjacent rectangles which realize a paving of the plane. The rectangular cell is not a primitive-cell: the nodes generated by Eq. (2.8) coincide with half the nodes of the Bravais lattice (the centers of the cells are not accounted for). Each cell has a surface twice that of a primitive unit-cell, and is in correspondance with two nodes of the Bravais lattice.

One could define other rectangular cells by taking multiples of $\vec{c_1}$ and $\vec{c_2}$ as edges, with however the drawback of generating an even smaller fraction of the Bravais lattice nodes. The rectangular cell defined by (2.8), which is the *conventional unit-cell* of the considered two dimensional lattice is the smallest of the rectangular cells.

More generally, the *conventional cell* of a crystal is the smallest parallelepiped, based on translations of the form (2.1), the shape of which displays in an obvious manner the symmetries (mirror reflexions, rotations) of the Bravais lattice. Figure 2.10 shows an example of conventional cell for three-dimensional Bravais lattices. In general, the conventional cell is a *multiple-cell*, associated to a density of several nodes per cell. There are cases, however, in which the conventional cell coincides with a primitive unit-cell. Such a situation defines *simple Bravais lattices*.

2.4.3   *Classification of the Bravais lattices. Cubic lattices*

The distinction between primitive cells and conventional ones provides a means for classifying Bravais lattices. The resulting classification is a two-level one.

The first level consists in grouping together all the Bravais lattices possessing conventional cells having the *same shape*. For instance, a Bravais lattice will be classified as *cubic* if its conventional cell is a *cube*. For a cube of edge $\mathbf{c}$, the translations $(\vec{c_1}, \vec{c_2}, \vec{c_3})$ defining the conventional cell satisfy the conditions:

$$|\vec{c_1}| = |\vec{c_2}| = |\vec{c_3}| \qquad \vec{c_i}.\vec{c_j} = \mathbf{c}^2.\delta_{ij} \tag{2.9}$$

There are seven different shapes of conventional cells determining seven classes of lattices (Fig. 2.11) : cube (*cubic lattices*), prism with square basis (*tetragonal lattices*), prism with rectangular basis (*orthorhombic lattices*), prism with parallelogram basis (*monoclinic lattices*), prism with a 120 degrees diamond basis (*hexagonal lattices*), rhombohedron (*rhombohedral lattices*), parallelepiped with arbitrary shape (*triclinic lattices*).

Within each of the former classes, a second level of classification distinguishes *Bravais lattices types* on the basis of the form of the relationship which exists between the vectors defining respectively the conventional and the primitive cells.

Let us consider for instance the case of cubic lattices. Three types of such lattices can be distinguished[5] (Fig. 2.10)

---

[5]The distinction of only three distinct types of cubic Bravais lattices implies that we consider as equivalent the lattices having the same geometrical shape of conventional cell, but a different size (i.e. different value of $\mathbf{c}$).

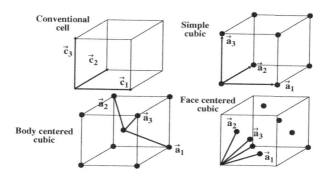

Figure 2.10   Conventional cubic cell and cubic, SC, BCC, and FCC primitive cells, The black circles are the nodes of the Bravais lattice.

a) Simple cubic lattice (abbreviated as SC)

For this lattice, the primitive and conventional cells coincide. The Bravais lattice is generated by the translations (2.9).

b) Body centered cubic lattice (abbreviated as BCC)

The primitive translations are related to those (2.9) defining the conventional cell by :

$$\vec{a_1} = \frac{1}{2}(\vec{c_1} + \vec{c_2} - \vec{c_3}) \quad \vec{a_2} = \frac{1}{2}(-\vec{c_1} + \vec{c_2} + \vec{c_3}) \quad \vec{a_3} = \frac{1}{2}(\vec{c_1} - \vec{c_2} + \vec{c_3})$$
(2.10)

The mixed product $(\vec{a_1}, \vec{a_2}, \vec{a_3})$, determining the volume of the primitive unit-cell, is half the volume $(\vec{c_1}, \vec{c_2}, \vec{c_3}) = \mathbf{c}^3$ of the conventional unit-cell. This result is in agreement with the fact that the nodes of the Bravais lattice occupy the vertices *and* the center of the conventional cubic cell (thus justifying the terminology used). The shortest distance between nodes, equal to the length of a primitive translation, is $d_{node} = |\vec{a_i}| = \frac{1}{2}\mathbf{c}\sqrt{3}$.

c) Face centered cubic lattice (abbreviated as FCC)

The relevant relationship is :

$$\vec{a_1} = \frac{1}{2}(\vec{c_1} + \vec{c_2}) \qquad \vec{a_2} = \frac{1}{2}(\vec{c_2} + \vec{c_3}) \qquad \vec{a_3} = \frac{1}{2}(\vec{c_3} + \vec{c_1})$$
(2.11)

The volume of the primitive unit-cell, $\frac{1}{4}\mathbf{c}^3$, is one fourth of that of the conventional cell, in agreement with the fact that the nodes of the Bravais lattice occupy he vertices *and* the face-centers of the conventional cubic cell (consistently with the terminology). The shortest distance between nodes is $d_{node} = |\vec{a_i}| = \frac{1}{\sqrt{2}}\mathbf{c}$.

If one performs for the other shapes of conventional cells the same enumeration of distinct lattice types, one finds that there are **14** *Bravais lattice types* unequally distributed among the seven shapes of conventional cells. Their traditional labelling is indicated on Fig. 2.11. Seven of these Bravais lattice types are *simple* (i.e. the primitive cell is identical to the conventional one).

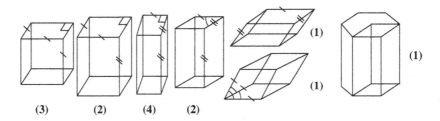

(1)

(1)

(1)

(3)   **(2)**   **(4)**   **(2)**

Figure 2.11   Seven cell shapes which are associated to the 14 Bravais lattice types. The numbers between brackets indicate the number of unequivalent lattices for each shape (and lattice symmetry). From left to right are represented the cubic, tetragonal, orthorhombic, monoclinic, triclinic, rhombohedral and hexagonal cell-shapes.

2.5   Examples of crystal structures

Some of the mechanical properties of solids can be explained by relying on a very simplified description of their atomic structure restricted to the mere specification of their Bravais lattice. However, a more detailed account of these properties (e.g. the specific properties of a solid of given composition) requires the knowledge of its complete crystal structure, i.e. of the chemical nature and position of all the atoms in its primitive unit-cell.

2.5.1   *Simple monoatomic structure packings*

One can expect that solids made from a single type of atom, as for instance pure metals (iron, copper...) will possess a simpler crystal structure than other solids. This is only partly true. It is generally observed that even a single chemical element will give rise, as a function of temperature or pressure, to several different crystalline structures some of which are fairly complex. This is even more the case for alloys made of several different atoms. Nevertheless, in a large class of monoatomic solids, mainly those made from elements situated in the left part of the Mendeleiev table (Fig. 2.17), the predominantly observed structure resembles the one obtained by packing together *hard spherical balls,* i.e. undeformable balls, in contact with eachother and whose volumes do not overlap.

There are several manners of packing hard spheres to build a crystalline structure. Two of the resulting structures are termed *closed-packed.* This qualification means that, for a given radius of the atomic sphere, the compacity of the packing is maximum : the voids existing between the atoms occupy a smaller fraction of the total volume than in any other packing.

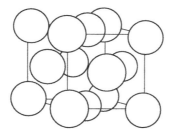

Figure 2.12   Atomic positions in the conventional cell of the compact cubic packing (FCC).

Cubic close-packing

This packing is obtained by placing an atom of the considered chemical species at each node of a FCC-lattice (Fig. 2.10). The edge-length **c** of the conventional-cubic cell is such as to ensure the contact between the atomic spheres. If **r** is the atomic radius, and $|a_i|$ the length of a primitive translation of the lattice, the latter condition implies the relationship $|a_i| = $ $\mathbf{c}/\sqrt{2} = 2\mathbf{r}$. Each atom is in contact with 12 closest neighbours situated along the face diagonals at a distance $2\mathbf{r} = |a_i|$.

The construction of the structure (one atom at each node) determines a density of one atom per primitive unit-cell. The conventional cell contains 4 atoms. The compacity of the structure can be characterized by the ratio between the volume of an atom and that of the primitive unit cell :

$$\frac{V_{atom}}{V_{cell}} = \frac{4\pi r^3/3}{\mathbf{c}^3/4} = \frac{\pi}{3\sqrt{2}} = 0.74 \tag{2.12}$$

Hexagonal close-packing

The construction of this packing is more complex (Fig. 2.13). Consider first
the hexagonal Bravais lattice (Fig. 2.11) generated by the set of primitive
translations $(\vec{a_1}, \vec{a_2}, \vec{c})$ where $\vec{a_1}$ and $\vec{a_2}$ are vectors of equal length making
an angle of 60 degrees, and where $\vec{c}$ is perpendicular to the $(\vec{a_1}, \vec{a_2})$ plane.
In each primitive cell, place one atom at the origin O and the other at O'
such as $\overrightarrow{OO'} = \frac{1}{3}(\vec{a_1} + \vec{a_2} + \vec{c}/2)$. The atoms of the structure will be in
direct contact with their neighbours if the values of the atomic radius **r** and
of the parameters **c** and **a** are related by :

$$\frac{\mathbf{c}}{\mathbf{a}} = \frac{\mathbf{c}}{2\mathbf{r}} = 2\sqrt{\frac{2}{3}} \tag{2.13}$$

In this structure, the basis contains *two* identical atoms (located respectively at O and at O'). Thus, the compacity is given by :

$$\frac{2V_{atom}}{V_{cell}} = \frac{8\pi r^3/3}{\mathbf{a}^2\mathbf{c}\sqrt{\frac{3}{2}}} = 0.74 \tag{2.14}$$

It has the same value as in the cubic close packing.

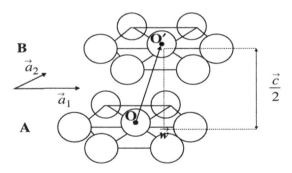

Figure 2.13   Atomic positions in the HC hexagonal compact structure.

If one applies to the atom located in O the lattice translations gener-
ated by $(\vec{a_1}, \vec{a_2}, \vec{c})$, one obtains a succession of planes (which we label **A**),
parallel to $(\vec{a_1}, \vec{a_2})$, spaced by $|\vec{c}|$, and containing one half of the atoms
of the structure. In each of these planes, atoms form a regular lattice of
equilateral triangles (Fig. 2.14). Clearly, this configuration is the most compact that can be formed in two dimensions with identical "hard disks" in

contact with each other. Note that the atoms of consecutive **A**-planes form columns parallel to $\vec{c}$.

Applied to the atom located in O', the set of lattice translations generates a second family of atomic-planes (labelled **B**) having identical internal configurations and spacing as **A**, each **B**-plane being at half distance between two **A**-planes. With respect to the columns generated by the **A**-atoms, the **B**-columns are shifted laterally by:

$$\vec{w} = \frac{1}{3}(\vec{a_1} + \vec{a_2}) \tag{2.15}$$

The **B**-atoms lye above or below voids between adjacent **A**-atoms.

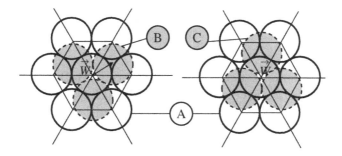

Figure 2.14   Structure of the most dense planes of the hexagonal compact and of the FCC structures. Relative atomic positions of adjacent A and B planes (Left) and of adjacent A and C planes.

The hexagonal close-packing can therefore be described as a stacking **A-B-A-B...** of planes having each a close-packed atomic configuration in two-dimensions, consecutive planes being shifted by $\pm\vec{w}$ with respect to each other.

Relationship between close-packings

In the cubic close-packing, the atomic planes perpendicular to one of the cube-diagonals, and defined by face-diagonals of the cubic cell, are the most dense lattice planes of the structure since they correspond to the shortest distances $|\vec{a_i}| = 2\mathbf{r}$ between atoms. It is easy to see that their structure is the close-packed two dimensional triangular configuration of atoms (Fig. 2.14). Similarly to the hexagonal close packing structure, the cubic close-packed one can be described as a stacking of close-packed planes.

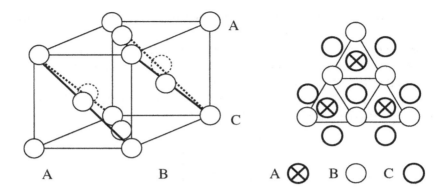

Figure 2.15   Decomposition of the FCC-compact structure into a stacking of hexagonal compact planes.

Figure 2.15 shows that the stacking sequence **A-B-C-A-B-C...** is different from the hexagonal sequence **A-B-A**. Indeed, along a cube-diagonal one finds successively : a) a plane **(A)** containing the lower corner of the cube. b) a plane **(B)** containing the diagonal of the lower face of the cube, with atoms shifted by the displacement $\overrightarrow{w}$ with respect to **A**. c) a plane **(C)** containing the diagonal of the upper face, with atoms shifted by $2\overrightarrow{w}$ with respect to **A**. The following plane, containing an atom at the upper corner, is an **A**-plane.

Hence, it is the shift between the second and third planes ($\mp\overrightarrow{w}$) which distinguishes the structures of the cubic and of the hexagonal close-packings. The similarity of construction of the two structures explains their identical compacity.

Body centered cubic packing

This packing (Fig. 2.16) is obtained by putting an atom at each node of a BCC Bravais lattice. The direct contact between nearest neighbour atoms requires that the radius of an atom equals half the length of the primitive translation of the lattice : $\mathbf{r} = |\overrightarrow{a_i}|/2 = \mathbf{c}\frac{\sqrt{3}}{4}$. The compacity of this packing is:

$$\frac{V_{atom}}{V_{cell}} = \frac{4\pi r^3/3}{\mathbf{c}^3/2} = 0.68 \tag{2.16}$$

It is smaller than in the close-packed structures. The shortest distance between atoms being the half-length of the cube-diagonal, the most dense planes are defined by two cube-diagonals (or two face-diagonals belonging to opposite faces). The structure of these planes is different from that of the close-packed planes, each atomic sphere being in contact with 4 atoms only instead of six.

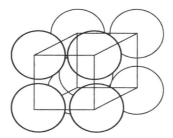

Figure 2.16   Monoatomic body centered cubic (BCC) structure.

2.5.2   *Physical realizations in metals*

Part of the Mendeleiev table is reproduced below (Fig. 2.17). All the selected elements are metals in the solid state, and give rise to structures which can be considered as realizations of the packings described above. Hence, a simple model of hard spheres in contact with eachother accounts satisfactorily for the structures actually observed in nature for a large number of pure metals.

This situation is in particular that of monovalent metals such as sodium. In this case the compact packing results from the fact that the binding forces between atoms are determined by the "free-electron gas" of valence-electrons of the atoms. This "gas" induces an effective attraction between the constituants which is approximately *isotropic* in space. As a consequence, there is no favoured direction of contact between atoms. The most stable structure will be the most compact one, which brings together in contact the largest number of atoms. We will see in chapter 7, that this "isotropy" of the metallic bond has also an important role in the easiness with which the plastic deformation can occur.

Such an explanation would not be relevant for covalent-bonded atoms. In this case, the angles between bonds have definite values (e.g. the tetrahedral configuration of bonds in carbon compounds). For most metals the bonding

									Al
Ti	V	Cr	Mn	Fe	Co	Ni	Cu	Zn	
Zr	Nb	Mo		Ru	Rh	Pd	Ag		
Hf	Ta	W		Os	Ir	Pt	Au		

Figure 2.17   Portion of the periodic table of elements. Black background: metals with a hexagonal compact structure. Grey background: BCC structure. White background: FCC structure.

is partly metallic and partly covalent. Nevertheless, quantum mechanical studies of the stability of metals show that even in this situation, each atom has a tendency to be surrounded by the largest possible number of atoms, thus justifying, also here, the stability of compact structures.

Metallic alloys

Metallic alloys, which are of major technological importance, are solids formed by several chemical elements. It is alloys, rather than pure metals, which provide the mechanical properties required by industrial applications.

The spatial configuration of their constituting atoms involves several complications. It can be *ordered* or *disordered*. This means that the periodicity of the crystal structure is, respectively, strictly, or statistically, realized. In the latter case, an atomic position is occupied randomly by one of the chemical elements of the alloy. On the other hand, an alloy can be heterogeneous and involve several crystal structures "mixed" at the micronic or submicronic scale. The composition of the alloy is then an average of that of the different structures. Let us give two examples of alloys, namely that of austenitic steel, an alloy of iron and carbon, and that of brass, which is an alloy of copper and zinc.

In a first approximation, the structure of austenitic steel is a compact cubic (FCC) packing of iron atoms. As indicated on the table in Fig. 2.17, this structure is not the equilibrium structure of iron at room temperature which is of the BCC type. Pure iron crystallizes in the FCC structure between 900°C and 1400°C. However, this structure can be stabilized at room temperature by the addition of a small proportion (less than 7%)

of carbon atoms.   These atoms partly occupy the so-called "octahedral"

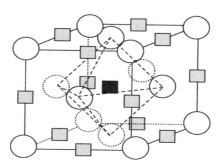

Figure 2.18   Octahedral sites (voids), represented by squares, within the atomic structure of an FCC monoatomic metal. The black shaded site at the center of the cube is surrounded by 6 atoms forming an octahedron.

voids of the FCC structure. As shown on Fig. 2.18, these voids are located at the center of the cubic cell, and at the center of its edges (these are "equivalent" to the center of the cell since they are related geometrically to it by primitive translations). Each such site is surrounded by six iron atoms forming a regular octahedron, thus justifying their name. The cubic cell, which contains four iron atoms also contains four such sites. The proportion of carbon atoms is always small. Certain cubic cells will contain a carbon atom while others will not contain any. Experimental studies reveal that the distribution of carbon atoms is random in the octahedral sites. In the ideal cubic compact structure, the size of an octahedral void can contain an atom of maximum radius $\rho = (0,414.r) = (0,146.c)$, where $r \cong 1$ Å, is the radius of the iron atom and c the edge of the cubic cell. Hence, $\rho \cong 0,42$ Å. The radius of a carbon atom is $\rho_c \cong 0,77$ Å. Its introduction in the structure will therefore deform significantly the octahedral site. This explains that the proportion of carbon atoms in the iron structure remains small.

Brass, of chemical composition CuZn, crystallizes in two types of structures labelled β and β' (Fig. 2.19). The latter is stable below 350°C and the former above this temperature. Both can be described by referring to cubic Bravais lattice in which the vertices and the center of the cubic cell are occupied by atoms. In the β structure, each site is occupied randomly by a copper or a zinc atom. Statistically, the vertices and the centers of the cubic cells are therefore equivalent. These two sets of points form together

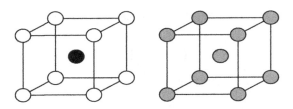

Figure 2.19   Structures of the β high-temperature (Right) and β' low-temperature (Left) phases of the copper-zinc alloy CuZn. In both cases the represented atomic sites are randomly occupied by copper or zinc atoms.

a centered cubic Bravais lattice. The basis of this "random" structure is then an "average" atom having probabilities of 50% of being a copper atom or a zinc atom.

In the β' structure, the centers of the cells are occupied by A atoms which have a probability p of being copper and $(1-p)$ of being zinc, while the vertices are occupied by B atoms having inverted proportions of being copper and zinc. In that case the vertices and the centers of a cell are not equivalent. The Bravais lattice of this random crystal is then of the simple cubic type, the basis being formed by a pair of atoms A and B. The probability p increases progressively on cooling below 350°C, to reach the value $p = 1$, at very low temperatures. Its only in this situation that the crystal is not random.

2.5.3   *Simple covalent structures*

The chemical elements situated in the right-hand part of the Mendeleiev table do not crystallize as simple atomic packings of hard spheres. Similarly to the case of solid-carbon, this is related to the fact that the cohesion between constituents of the corresponding solids relies on a covalent bonding having a marked directional character. Hence, an atom of silicon (Si) or of germanium (Ge) tends to be surrounded by four atoms forming a regular tetrahedron.

The crystal structure common to the three preceding chemical elements is the so-called "diamond structure". Its Bravais lattice is of the FCC type (2.12). Its basis contains two identical atoms (Fig. 2.20), and the conventional cell therefore contains 8 atoms. One is at the origin of the primitive unit cell $(\vec{a_1}, \vec{a_2}, \vec{a_3})$ defined by Eq. (2.11). The second is located at $\frac{1}{4}(\vec{a_1}+\vec{a_2}+\vec{a_3}) = \frac{1}{4}(\vec{c_1}+\vec{c_2}+\vec{c_3})$. This structure optimizes the configuration

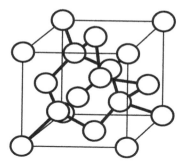

Figure 2.20   Structure of diamond. Each carbon atom is at the center of a regular tetrahedron formed by four other carbon atoms.

of the bonds of each atom (C, or Si, or Ge) since each atom is at the center of a tetrahedron formed by its closest neighbours. If we assume the atoms to be spheres in contact with their neighbours, the relationship between their radius $\mathbf{r}$ and the cube-edge $\mathbf{c}$ is $\mathbf{r} = \frac{1}{8}|\vec{c_1} + \vec{c_2} + \vec{c_3}| = \mathbf{c}\frac{\sqrt{3}}{8}$. The compacity of the structure can then be evaluated as :

$$\frac{8V_{atom}}{V_{cell}} = \frac{32\pi\mathbf{r}^3/3}{\mathbf{c}^3} = \frac{\pi\sqrt{3}}{16} = 0.34 \qquad (2.17)$$

It is much smaller than that of the closed-packed structures (0.74). The strong cohesion of the diamond structure, which is reflected, for instance, in the mechanical hardness of the crystals, is not related to the compacity of the atomic configuration but, rather, to the strength of the covalent bonds.

2.6   Non-crystalline solids

The description of crystals has been focused on their geometrical characteristics. An alternate point of view consists in considering a crystal as the most *ordered* state of its microscopic constituants. This concept pertains to the *statistical* description of a macroscopic system. For such a system, the large number of the constituants (of the order of the Avogadro number) is a source of uncertainty on their state, the magnitude of this uncertainty being measured by the *statistical entropy*. We are concerned with the *positional uncertainty* of the large assembly of atoms constituting a crystal. In this aspect, a system will be considered as perfectly ordered, if its positional *state* (i.e. the positions of all its atoms), can be *entirely specified*. *The po-*

sitional entropy of the crystal has zero value, no uncertainty being involved in the determination of the positions of the atoms. A crystal at very low temperatures ("thermal defects" are then absent), can be considered as an ideally-ordered system. Indeed, one can define the set of positions of all the atoms in the structure by specifying the positions of the atoms composing the sole *basis*, and the coordinates of the three primitive translations. In this crystalline order, one can distinguish two parts. On the one hand, a crystal has *translational order*, residing in the existence of a Bravais lattice underlying the structure. On the other hand, it has *orientational order*, residing in the definite orientation in space of the group of atoms located at each node of the lattice.

Among solids, *glasses* constitute, as opposed to crystals, a *reference of statistical disorder*. The translational order is missing, the distances between the different groups of atoms being statistically uncorrelated beyond a few angströms (no Bravais lattice). The orientational order is also missing, the orientations of atomic groups being random and uncorrelated beyond a distance of a few angströms (there is however a short-range correlation imposed by the constrained orientations of the chemical bonds between constituents). Ordinary glass is mainly based on "silicates" in which the cluster SiO_2 plays an important role in the binding of the constituants.

Another interesting case of non-crystalline solids is provided by *quasicrystals.*Their structure has an underlying lattice which can be constructed by using *two distinct shapes of adjacent cells*, instead of a single one for ordinary Bravais lattices. The shapes and sizes of these cells are such as to permit, through definite assembling rules, a paving of the three-dimensional space by adjacent cells. An illustration, in two-dimensions, is provided by Fig. 2.21.

Such a lattice has no periodicity since any two nodes possess a different surrounding. However, if attention is given to the immediate surrounding (the nearest neighbours) of nodes, it is found that an infinity of nodes are equivalent.

The structural differences between crystalline and non-crystalline materials are central to the explanation of their plastic properties. We will see that the mobility of dislocations, which are a condition of the plastic deformation, is realized in many crystalline metals and is not realized in non-crystalline materials thus explaining the brittleness of the latter materials.

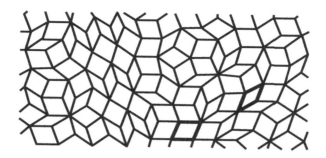

Figure 2.21   Two-dimensional model of paving by a quasi-crystalline set of cells having one of two shapes (elongated diamond and wide diamond).

Chapter 3

Mechanics of deformable solids

Main concepts: Definition of the fundamental tensors. Symmetries of the elastic tensor. Case of an isotropic solid. Main formulas.

3.1 Introduction

This chapter is devoted to a brief recall of the elementary concepts and formulas relative to the linear elastic properties of solids. Although the theory contained in the following chapters is essentially developped for *isotropic* solids, some indications are given here on the form of the stiffness tensor in crystalline solids. This will justify the use, in the case of isotropic media, of two independent stiffness coefficients only. These indications will also facilitate the access to more specialized treatises in which the effects of the crystal anisotropy on plastic properties are examined.

3.2 Fundamental tensors

3.2.1 *Strain and stress*

The mechanical state of a continuous solid, deformed with respect to a reference state of the solid, considered as non-deformed, can be characterized by a *displacement field* $\vec{\xi}(\vec{r})$ defined at each point of this medium. A uniform displacement will only translate the solid as a whole, and thus induce no deformation. A deformation is then necessarily associated to the existence of an *inhomogeneous* displacement, associated to a non-zero gradient of $\vec{\xi}(\vec{r})$. More precisely, at a given point of the solid, the local deformation is defined by the components of the so-called *strain tensor* whose expression for small values of the gradient of $\vec{\xi}(\vec{r})$, has the form:[1,2]

$$\bar{\bar{\epsilon}} \rightarrow \epsilon_{ij} = \frac{1}{2}\left(\frac{\partial \xi_i}{\partial x_j} + \frac{\partial \xi_j}{\partial x_i}\right) \tag{3.1}$$

The physical meaning of each of the six components of this symmetric second rank tensor [3] is recalled in Fig. 3.1. The three diagonal components

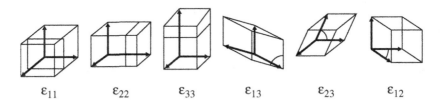

$$\varepsilon_{11} \qquad\qquad \varepsilon_{22} \qquad\qquad \varepsilon_{33} \qquad\qquad \varepsilon_{13} \qquad\qquad \varepsilon_{23} \qquad\qquad \varepsilon_{12}$$

Figure 3.1   Components of the strain tensor. The ϵ_{ii} measure the relative expansion along the three axes of the reference frame, and the ϵ_{ij} the change of angles (with respect to the right angle) between the axes.

ϵ_{ii} are the values of the relative stretching of the solid along the three axes of the rectangular frame of coordinates. If the solid is locally compressed along one of the axes, the corresponding value ϵ_{ii} has a negative value. Their sum $\delta = \sum \epsilon_{ii} = div\,\vec{\xi}$ is equal to the local change in volume $(\delta\Omega/\Omega)$. The three components ϵ_{ij} $(i \neq j)$, called the *shear components* express the variations of the angles between the frame axes.[4]

The internal stress, which defines the mutual forces exerted between adjacent regions of a continuous medium, is also represented mathematically by a symmetric second rank tensor σ_{kl}. Thus, the force $d\vec{f}_{21}$ exerted across a small surface dS by region (2) of a solid, on region (1) is written as (Fig. 3.2):

$$d\vec{f}_{21} = \bar{\bar{\sigma}}.\vec{n}_{12}dS \tag{3.2}$$

where $\vec{n}_{12}$ is the unit vector perpendicular to the surface dS, oriented from region (1) towards region (2). The three "longitudinal" components

[1]For larger gradients the relationship is: $\epsilon_{ij} = \frac{1}{2}\left(\frac{\partial \xi_i}{\partial x_j} + \frac{\partial \xi_j}{\partial x_i} + \frac{\partial \xi_i}{\partial x_j}.\frac{\partial \xi_j}{\partial x_i}\right)$

[2]The antisymmetric part of the gradient of $\vec{\xi}$ défines the infinitesimal rotation of a local rectangular frame of coordinates.

[3]The rank of a tensor is equal to the number of indices used to label the components.

[4]If the angle between the axes i and j, initially rectangular, becomes $(\frac{\pi}{2} - \delta\theta_{ij})$ with $\delta\theta_{ij}$ small, one has $\epsilon_{ij} = \delta\theta_{ij}/2$. Sometimes the *shear angle* is identified with the angle $\gamma = \delta\theta_{ij}$ equal to twice the value of ϵ_{ij}.

σ_{ii} represent forces of *traction* (or stretching) in the direction of the three axes of coordinates, while the σ_{ij} ($i \neq j$) are shear forces. They tend to "close" the angles between the coordinate axes (Fig. 3.1). A hydrostatic pressure p corresponds (with an inversion in sign) to a scalar stress tensor (i.e. having three equal diagonal components and non-diagonal ones all equal to zero):

$$\sigma_{kl} = -p\delta_{kl} \tag{3.3}$$

The mechanical equilibrium of a small cube of matter centered on point $M(\overrightarrow{r})$ within the solid, and submitted to the forces of the surrounding regions, such as the one represented on Fig. 3.2, is expressed by the relationship:

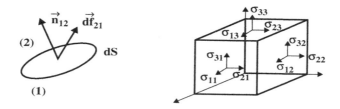

Figure 3.2 (Left) Definition scheme of the stress tensor. (Right) Direction of the forces associated to the different components of the stress tensor. These forces are exerted by the external medium on the matter contained within a cube across its faces.

$$\sum_k \frac{\partial \sigma_{kl}}{\partial x_k} = 0 \tag{3.4}$$

On the other hand, *mechanical energy* is stored locally in a solid, wherever non-zero tensor fields of strain $\epsilon_{ij}(\overrightarrow{r})$ and stress $\sigma_{ij}(\overrightarrow{r})$ exist. Its density $u(\overrightarrow{r}) = (dU/d\Omega)$ in a small volume surrounding point $M(\overrightarrow{r})$ is:

$$u(\overrightarrow{r}) = \frac{dU}{d\Omega}(\overrightarrow{r}) = \frac{1}{2}\sum_{ij} \sigma_{ij}(\overrightarrow{r})\epsilon_{ij}(\overrightarrow{r}) \tag{3.5}$$

3.2.2 *Stiffness*

Among solids, *linearly elastic solids* are characterized by a linear relationship between the components of the strain tensor and those of the stress tensor. This relationship expresses Hooke's law:

$$\sigma_{ij} = \sum_{kl} C_{ijkl} \cdot \epsilon_{kl} \tag{3.6}$$

where $\sigma_{ij}, \epsilon_{kl}$, and C_{ijkl} are respectively the cartesian components of the stress tensor, the strain tensor, and the fourth rank tensor of *elastic stiffness* $\overline{\overline{\overline{C}}}$.

In mechanics, this law can be considered as empirical. The number and values of the elastic stiffness coefficients C_{ijkl} are experimental data which are determined by various methods of measurement. Large values of the C_{ijkl} mean that the solid material is "stiffer", i.e. less deformable for given applied forces. The linear relationship defining Hooke's law can also be expressed by the "inverse" relationship between $\overline{\overline{\epsilon}}$ and $\overline{\overline{\sigma}}$:

$$\epsilon_{ij} = \sum_{kl} S_{ijkl} \sigma_{kl} \tag{3.7}$$

where $\overline{\overline{\overline{S}}}$ is called the *elastic compliance* tensor. Large values of the coefficients S_{ijkl} mean that the solid is more easily deformable.

From Eqs. (3.6), and (3.5), one can draw the following expression for the mechanical energy density:

$$u = \frac{1}{2} \sum_{ijkl} C_{ijkl} \epsilon_{ij} \epsilon_{kl} \tag{3.8}$$

3.3   Coordinate changes

In the coming chapters, one will consider several microscopic objects (vacancies, dislocations) whose specific geometrical shape will impose to adopt a spherical or a cylindrical system of coordinates (Fig. 3.3). In such frames the tensor components can be deduced from the cartesian ones through a transformation rule discussed in the next paragraph.

Table 3.1 Cartesian components T_{ij} of a second rank tensor and components in the "cyindrical" frame of coordinates.

T_{rr}	$T_{11}\cos^2\theta + 2T_{12}\sin\theta\cos\theta + T_{22}\sin^2\theta$	T_{rz}	$T_{13}\cos\theta + T_{23}\sin\theta$
$T_{\theta\theta}$	$T_{11}\sin^2\theta - 2T_{12}\sin\theta\cos\theta + T_{22}\cos^2\theta$	$T_{\theta z}$	$-T_{13}\sin\theta + T_{23}\cos\theta$
$T_{r\theta}$	$-T_{11}\sin\theta\cos\theta + T_{12}\cos 2\theta + T_{22}\sin\theta\cos\theta$	T_{zz}	T_{33}

Tables 3.1 and 3.2, are deduced by application of this rule. They show, for second rank symmetric tensor $\overline{\overline{T}}$ (e.g. the strain or stress tensor), the correspondance between the cartesian components and, respectively the cylindrical and spherical coordinates. Note that the definition (3.1) of the

strain tensor is only valid for a cartesian frame of reference with fixed orientation (i.e. one that does not change with the current point $\vec{r}$). This is not the case of the frames used for cylindrical or spherical coordinates, since these are local rectangular frames which depend on $\vec{r}$.

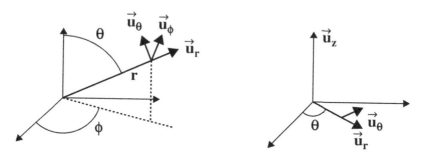

Figure 3.3   Local frames adapted to spherical coordinates (Left) and to cylindrical co-ordinates (Right).

Hence, in order to express the "spherical" or "cylindrical" components of these tensors, one has to compute the cartesian components (3.1) of these tensors and use the conversion Tables 3.1 and 3.2.

Table 3.2 Cartesian components T_{ij} of a Second rank tensor and its components in spherical coordinates.

$\mathbf{T}_{rr}$	$T_{11}\sin^2\theta\cos^2\phi + T_{22}\sin^2\theta\sin^2\phi + T_{33}\cos^2\theta+$ $2T_{12}\sin^2\theta\sin\phi\cos\phi + 2T_{13}\sin\theta\cos\theta\cos\phi + 2T_{23}\sin\theta\cos\theta\sin\phi$
$\mathbf{T}_{\phi\phi}$	$T_{11}\sin^2\phi + T_{22}\cos^2\phi - 2T_{12}\sin\phi\cos\phi$
$\mathbf{T}_{\theta\theta}$	$T_{11}\cos^2\theta\cos^2\phi + T_{22}\cos^2\theta\sin^2\phi + T_{33}\sin^2\theta+$ $2T_{12}\cos^2\theta\sin\phi\cos\phi - 2T_{23}\sin\theta\cos\theta\sin\phi - 2T_{13}\sin\theta\cos\theta\cos\phi$
$T_{r\phi}$	$-T_{11}\sin\theta\sin\phi\cos\phi + T_{22}\sin\theta\sin\phi\cos\phi+$ $T_{12}\sin\theta\cos 2\phi + T_{23}\cos\theta\cos\phi - T_{13}\cos\theta\sin\phi$
$T_{r\theta}$	$-T_{11}\sin\theta\cos\theta\cos^2\phi - T_{22}\sin\theta\cos\theta\sin^2\phi + T_{33}\sin\theta\cos\theta$ $-2T_{12}\sin\theta\cos\theta\sin\phi\cos\phi - T_{23}\cos 2\theta\cos\phi - T_{13}\cos 2\theta\sin\phi$
$T_{\theta\phi}$	$(T_{11} - T_{22})\cos\theta\sin\phi\cos\phi$ $-T_{12}\cos\theta\cos 2\phi - T_{23}\sin\theta\sin\phi + T_{13}\sin\theta\cos\phi$

3.4 Stiffness tensor and crystal symmetry

3.4.1 *General constraints*

The tensor representing the elastic stiffness has $(3)^4 = 81$ components. However, certain components are linearly related to eachother. Part of these relationships are valid for all solids. Others depend of the crystal symmetry and are common to all the solids with the same crystal symmetry. They define the *form* of the C_{ijkl} tensor. As a result, the number of *independent* (unrelated) components C_{ijkl} is smaller than 81. Let us analyze the various constraints imposed to the stiffness coefficients.

1°) For the symmetric strain and stress tensors, one has $\sigma_{ij} = \sigma_{ji}$ and $\epsilon_{kl} = \epsilon_{lk}$. Each of these tensors has 6 distinct components. Using these equalities, and the linear relationship 3.6, valid for any values of the ϵ_{kl}, the following equalities between stiffness coefficients hold for any solid material:

$$C_{ijkl} = C_{jikl} = C_{ijlk} = C_{jilk} \tag{3.9}$$

Hence, there are, at most, 36 (i.e. 6x6) distinct stiffness coefficients C_{ijkl}. It is customary to relabel the four indices $(ijkl)$ in the view of forming a 6x6 "stiffness matrix". One uses the correspondance $C_{ijkl} \to C'_{\alpha\beta}$ with $(ii) \to i; (23) \to 4; (13) \to 5; (12) \to 6$. This relabelling of indices can also be used for the six distinct stress or strain tensor coefficients. However, in this case, in order to preserve the form of Hooke's law (3.6) as well as that of the elastic energy (3.8), one puts $\epsilon_{ii} = \epsilon_i$ and $\epsilon_\alpha(\alpha > 3) = 2\epsilon_{ij}(i \neq j)^5$. The expression of the mechanical energy density, and the Hooke's law can then be written:

$$u = \frac{1}{2} \sum_{\alpha\beta} C'_{\alpha\beta} \epsilon_\alpha \epsilon_\beta \tag{3.10}$$

and

$$\sigma_\alpha = \sum_\beta C'_{\alpha\beta} \epsilon_\beta \tag{3.11}$$

2°) A second general relationship derives from consideration of Eqs. (3.6) and (3.8):

$$C_{ijkl} = \frac{\partial \sigma_{ij}}{\partial \epsilon_{kl}} = \frac{\partial^2 u}{\partial \epsilon_{ij}\partial \epsilon_{kl}} = \frac{\partial^2 u}{\partial \epsilon_{kl}\partial \epsilon_{ij}} = \frac{\partial \sigma_{kl}}{\partial \epsilon_{ij}} = C_{klij} \tag{3.12}$$

---

[5] Without this factor the summation over the indices α and β would involve factors of 2 in front of some terms of the sum. Note that this convention does not hold for the stress components for which for any α value, one has $\sigma_\alpha = \sigma_{ij}$.

or:

$$C'_{\alpha\beta} = C'_{\beta\alpha} \tag{3.13}$$

Thus the stiffness matrix $C'_{\alpha\beta}$ is *symmetric*. This symmetry reduces, for all solids, the maximum number of distinct stiffness coefficients to 21.

3.4.2 Crystal symmetry

A third restrictive constraint derives from the specific crystal symmetry of the solid considered. Thus a metal with a cubic structure will not posses the same form of stiffness tensor C_{ijkl} as a metal with a hexagonal symmetry.

Indeed, the values of the C_{ijkl} depend of the frame of reference, similarly to the components of a vector. They are modified when a geometrical transformation (e.g. a rotation) is applied either to the frame, or to the solid. On the other hand, if the considered geometrical transformation is a symmetry of the crystal structure of the solid, the components must keep their initial values, since the structure is transformed into an indistinguishable one.

As indicated in chapter 2, the symmetries of a crystal can be noted $\{S|\vec{t}\}$, and consist of a "point" transformation S (rotation, symmetry about a plane, etc...), followed by a translation $\vec{t}$, such as $|\vec{t}|$ is of the same order of magnitude as the distance between atoms (a few Angströms).

Let us first examine the action of $\vec{t}$. Note that the $C_{ijkl}(\vec{r})$ are defined for a solid considered as a continuous medium. Hence, the scale of variations of the tensor components is orders of magnitude larger than the atomic scale. In other terms, the values of the C_{ijkl}, must be considered as *constant* over distances $\Lambda \gg |\vec{t}|$. A displacement of the solid by $\vec{t}$ will therefore leave unchanged the values of the tensor components.

By contrast, a point transformation S will act, in general, on the tensor components, and, in particular on the C_{ijkl}. The nature of this action is detailed in the next paragraph.

In summary, the symmetry, at the atomic scale, of a crystalline solid, implies that the point transformations S must transform the components C_{ijkl} in such a way as to leave unchanged the form of the tensor $\overline{\overline{\overline{C}}}$.

3.4.3 Mathematical transformation of tensors

Let the three components of a vector $\vec{V}$ be denoted $\mathbf{v}_i$ ($i = 1, ..3$). Their transformation by a point transformation S is a well defined *linear transfor-*

mation. For instance, the symmetry about the origin will transform each $\mathbf{v}_i$ into $\mathbf{v}'_i = (-\mathbf{v}_i)$. More generally, for a rotation, or another point-symmetry S, one can write the action of S as a linear relationship:

$$\mathbf{v}'_i = \mathcal{S}(\mathbf{v}_j) = \sum_{\alpha=1}^{3} M_{i\alpha}\mathbf{v}_\alpha \tag{3.14}$$

By definition, the components of a fourth-rank Cartesian tensor C_{ijkl} are transformed, formally, by the action of S, in the same way as the product $\mathbf{v}_i\mathbf{v}_j\mathbf{v}_k\mathbf{v}_l$ of the components of a vector. Thus, if,

$$\mathcal{S}\mathbf{v}_i = \sum_\alpha M_{i\alpha}\mathbf{v}_\alpha \quad ; \mathcal{S}\mathbf{v}_j = \sum_\beta M_{j\beta}\mathbf{v}_\beta \tag{3.15}$$

$$\mathcal{S}\mathbf{v}_k = \sum_\gamma M_{j\gamma}\mathbf{v}_\gamma \quad ; \mathcal{S}\mathbf{v}_l = \sum_\delta M_{j\delta}\mathbf{v}_\delta \tag{3.16}$$

Then[6]:

$$\mathcal{S}(C_{ijkl}) = \sum_{\alpha,\beta,\gamma,\delta} M_{i\alpha}M_{j\beta}M_{k\gamma}M_{l\lambda}C_{\alpha\beta\gamma\lambda} \tag{3.17}$$

As an example, let us consider the case of a cubic crystal (this is the case of many usual metals such as iron, copper, etc...). Among the S point relevant operations, there are $2\pi/3$, and $4\pi/3$ rotations around the diagonals of the cubic cell. Such rotations will interchange the three coordinate axes. In application of the preceding rule (3.17), $\mathbf{C}_{1111}$ is therefore transformed into $\mathbf{C}_{2222}$ or into $\mathbf{C}_{3333}$, and conversely. The invariance of the tensor then implies that these coefficients must be equal:

$$\mathbf{C}_{1111} = \mathbf{C}_{2222} = \mathbf{C}_{3333} \tag{3.18}$$

The same argument leads to:

$$\mathbf{C}_{1122} = \mathbf{C}_{2233} = \mathbf{C}_{1133} \tag{3.19}$$

and:

$$\mathbf{C}_{1212} = \mathbf{C}_{2323} = \mathbf{C}_{1313} \tag{3.20}$$

Finally, if we consider a symmetry of the cube consisting in a reflexion about a plane passing by the center of the cell and perpendicular to the third coordinate axis we must have:

$$\mathbf{C}_{2213} = -\mathbf{C}_{2213} = 0 \tag{3.21}$$

---

[6] For instance, for a rotation of angle θ around the $\overrightarrow{Oz}$ axis, one has: $x'_1 = (x_1\cos\theta - x_2\sin\theta)$, $x'_2 = (x_1\sin\theta + x_2\cos\theta)$, $x'_3 = x_3$. The coefficients in the linear transformation are: $M_{11} = M_{22} = \cos\theta; M_{12} = -M_{21} = -\sin\theta; M_{13} = M_{31} = M_{23} = M_{32} = 0; M_{33} = 1$.

as well as similar relations. As a result, the 6x6 matrix C'_{ij} représenting the stiffness tensor has only 12 non-zero components among which three only are distinct : $C'_{11} = C_{1111}$; $C'_{12} = C_{1122}$; and, $C'_{44} = C_{1212}$. In all solid materials having a cubic conventional cell, the matrix of stiffness coefficients will have the form:

$$\overline{\overline{C}}_{cub} \rightarrow \begin{bmatrix} C'_{11} & C'_{12} & C'_{12} & 0 & 0 & 0 \\ C'_{12} & C'_{11} & C'_{12} & 0 & 0 & 0 \\ C'_{12} & C'_{12} & C'_{11} & 0 & 0 & 0 \\ 0 & 0 & 0 & C'_{44} & 0 & 0 \\ 0 & 0 & 0 & 0 & C'_{44} & 0 \\ 0 & 0 & 0 & 0 & 0 & C'_{44} \end{bmatrix} \qquad (3.22)$$

The following table indicates the values of the three distinct stiffness coefficients for some examples of well known metals having a cubic structure. Aluminum, copper and gold possess a compact FCC structure, while iron has a BCC structure.

Table 3.3 Stiffness coefficients for some cubic metals in units 10^{10} Pascals $\approx$ Tons/mm^2.

Material	C'_{11}	C'_{44}	C'_{12}
aluminum	11	2,8	6,3
copper	17	7,5	12
gold	19	4	16
iron	23	12	13,5
diamond	102	49	25

$$(3.23)$$

3.5   Isotropic solids

3.5.1   *Stiffness tensor*

A crystalline solid is never isotropic: the atomic arrangement as well as most physical properties depend on the direction in space. However, metals are often used in "polycrystal form". They are formed from the assembly of small crystalline grains of micronic size with random orientations. A metal sample with dimensions large with respect to the grain dimensions, can then be considered from the standpoint of its mechanical properties as an isotropic medium.

Such a medium will be invariant by a larger number of S point symmetry elements than cubic media. Hence, the relationships between stiffness coefficients of a cubic medium will hold, and additional relationships will be valid. Consider a rotation of angle $\pi/4$ around the z axis. One finds easily that this geometrical transformation, which does not respect the cubic symmetry, will impose the new relationship:

$$C'_{44} = \frac{1}{2}(C'_{11} - C'_{12}) \tag{3.24}$$

It can be shown that his additional relationship, is sufficient to ensure the invariance of the stiffness tensor with respect to the application of any symmetry of an isotropic medium. It has the effect of reducing to *two* the number of independent stiffness coefficients of an isotropic solid. It is customary to put $\mu = C'_{44}$ and $\lambda = C'_{12}$. λ and μ are called the *Lamé coefficients*. Expressed as function of the two coefficients, the matrix representing the stiffness tensor has the form:

$$\overline{\overline{C}}_{iso} \rightarrow \begin{bmatrix} \lambda+2\mu & \lambda & \lambda & 0 & 0 & 0 \\ \lambda & \lambda+2\mu & \lambda & 0 & 0 & 0 \\ \lambda & \lambda & \lambda+2\mu & 0 & 0 & 0 \\ 0 & 0 & 0 & \mu & 0 & 0 \\ 0 & 0 & 0 & 0 & \mu & 0 \\ 0 & 0 & 0 & 0 & 0 & \mu \end{bmatrix} \tag{3.25}$$

3.5.2 *Basic equations*

Replacing the stiffness coefficients C_{ijkl} as function of the Lamé coefficients, we can particularize the form of Eqs. (3.4), (3.6), and (3.8) for an isotropic medium. This form will be used extensively in the coming chapters.

Hence Hooke's law can be written as:[7]

$$\sigma_{kl} = \lambda\delta_{kl}.\left(\sum \epsilon_{ii}\right) + 2\mu\epsilon_{kl} \tag{3.26}$$

---

[7]If one uses the contracted form of strain components, the second term in Eq. (3.26) is $\mu\epsilon_\alpha$ or $\mu\gamma$ where γ is the shear angle.

On the other hand, the bulk compressibility χ of a solid is defined as the ratio betwwen the relative decrease of the volume $(-\sum \epsilon_{ii})$ and the pressure (3.3):

$$\chi = \frac{-1}{\Omega}\frac{\delta\Omega}{p} = \frac{-\sum \epsilon_{ii}}{-\sigma} = \frac{1}{(\lambda + 2\mu/3)} \tag{3.27}$$

Likewise, one deduces easily from (3.26) that, if a longitudinal traction σ_{ii} is exerted on an isotropic solid, the other stress components being set to zero, one obtains a relative extension-strain ϵ_{ii} equal to:

$$\sigma_{ii} = Y.\epsilon_{ii} = \frac{3\lambda + 2\mu}{\lambda + \mu}\mu.\epsilon_{ii} \tag{3.28}$$

Y is called the *Young modulus* of the solid.

In a linearly elastic solid, the density of elastic energy takes the form:

$$u = \frac{dU}{d\Omega} = \frac{\lambda}{2}(\sum_{i=1}^{3} \epsilon_{ii})^2 + \mu \sum_{i,j=1}^{3} \epsilon_{ij}^2 \tag{3.29}$$

Finally, substituting the displacements $\overrightarrow{\xi}(\overrightarrow{r})$ to the strain components, and using Eqs. (3.26) and (3.1), the elastic equilibrium of an isotropic solid will be defined by:

$$\mu\Delta\overrightarrow{\xi} + (\lambda + \mu)\overrightarrow{\nabla}(\overrightarrow{\nabla}.\overrightarrow{\xi}) = 0 \tag{3.30}$$

Alternately, introducing the dimensionless *Poisson coefficient* ν:

$$\nu = \frac{\lambda}{2(\lambda + \mu)} \tag{3.31}$$

the equilibrium equation takes the form:

$$(1 - 2\nu)\Delta\overrightarrow{\xi} + \overrightarrow{\nabla}(\overrightarrow{\nabla}.\overrightarrow{\xi}) = 0 \tag{3.32}$$

Chapter 4

Vacancies, an example of point defects in crystals

Main concepts: Dimensional classification of defects. Existence of stable defects at thermodynamic equilibrium. Formation and migration energies of point defects. Jump mechanism of vacancies. Random walk. Diffusion.

4.1 Classification of defects in crystals

A crystal defect consists in any deviation from the strict periodicity of the ideal (infinite) crystal described in chapter 2. Its relevance in a book devoted to the plastic behaviour of solids comes from the fact that it is the very existence of defects which explains the characteristics of this behaviour.[1]

The occurence of a defect always increases the *energy* of a crystalline solid. Indeed, the ideal periodic structure is believed to correspond to the lowest energy of the assembly of atoms constituting the solid. Consequently, at very low (absolute zero) temperature, the stable state of the solid, which then corresponds to its lowest energy, will be defect-free. However, at any other temperature, the thermodynamic equilibrium of a system is not determined by its sole energy, as will be recalled in section 4.2, and a real solid will differ from the ideal crystal model. *Certain types of defects* will be its *intrinsic* constituants. By contrast, other types of defects (in particular dislocations) should not be present at thermodynamic equilibrium. Their actual existence is due either to the out-of-equilibrium procedure of elaboration of the solid, or to the effect of specific external forces.

Defects can be classified according to their "dimensionality". Thus, one can consider *point defects* (zero-dimensional). For this type of defects, the volume of the perturbed region of a crystal is of the same order of

---

[1]Other important physical properties of solids are also determined by the occurence of defects, such as, for instance, the electrical resistivity of metals.

magnitude as the volume of a single atom, or of a few atoms. A simple example is the *vacancy* consisting in the absence of an atom in a site normally occupied by a constituent of the crystal (Fig. 4.1). Another simple example is the *interstitial*, i.e. an additional atom located in a normally empty space situated between the constituting-atoms of the structure (as the case of the carbon atoms in the iron structure described in chapter 2, section 4). A solid can also have *substitutional impurities*. These are atoms occupying a site of the crystal-structure normally occupied by an atom of different chemical nature. More complex point defects also exist, consisting of clusters (groups) of the preceding simple defects.

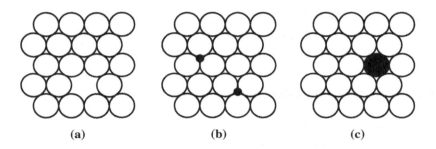

(a)   (b)   (c)

Figure 4.1   Three examples of point defects. (a) Vacancy. (b) Interstitial atom. (c) Substitutional atom.

A *linear defect* (one-dimensional) is a filamentary (thread-shaped) defect, such as, for instance, the absence of a row of atoms in a crystal. The section of such a defect has the same order of magnitude as the section of an atom or of a few atoms, while its length is large as compared to the atomic dimensions, and can be as large as the linear size of the macroscopic crystal sample.

This class of defects contains the *dislocations,* which, as mentioned in chapter 1, play a central role in the mechanism of plasticity. Their detailed description and properties will be analyzed in the next three chapters. In anticipation, Fig. 4.2 represents a simple type of dislocation, the so-called straight *edge* dislocation. This defect is associated to the absence a half plane of atoms in the structure. The dislocation is then the straight line limiting the missing half-plane. In the neigbourhood of this line, the configuration of the chemical bonds between atoms is strongly perturbed as compared to the normal configuration in the ideal crystal structure.

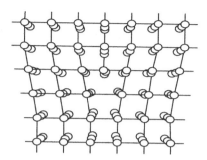

Figure 4.2 "Edge-type" dislocation. The upper part of the crystal contains an additional half atomic plane. The lower edge of this half-plane, and its neighbourhood, in which the atomic bonds are deformed, constitute a linear defect normal to the plane of the figure, called a "dislocation line".

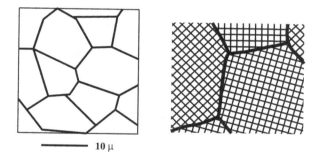

Figure 4.3 (Left) Schematic configuration of a polycrystalline sample containing adjacent grains of micronic size, separated by "grain boundaries". (Right) The grains have identical lattices which are differently oriented in space.

A *planar defect* (two dimensional) corresponds to a perturbed region of a crystal, whose volume is comparable to that of a plane of atoms, i.e. with a thickness comparable to a single atom or to a few atoms. Examples of such defects are the surfaces limiting externally a crystalline sample, or the "grain boundaries", which are the surfaces separating the adjacent crystalline grains of micronic dimension which constitute, ordinarily, metal samples.(Fig. 4.3). Other examples are the so-called "twin boundaries" separating two regions of a crystal having identical structures, differently oriented in space, and with a definite relative orientation (Figs. 4.4 (a), and 4.5). To this class of defects, also belong the specific "stacking faults" characterizing certain structures. Figure 4.4(b), represents such a defect in the case of the cubic compact structure described in chapter 2, section 4.

Various types of *three-dimensional defects* can be considered. For instance small volumes of a crystal phase distinct from the one considered can be scattered through the solid, and constitute "precipitates".

The above "dimensional" classification of defects has the advantage of being intuitive. However its consistency has sometimes to be questioned. Indeed, although a point-defect will only modify the configuration of chemical bonds in its immediate vicinity, it is also likely to induce a significant mechanical deformation at a distance of several tens of atomic intervals. It could therefore also be considered as a 3-dimensional defect.

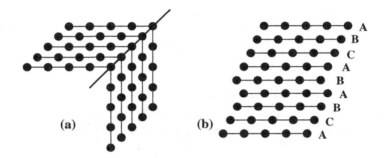

Figure 4.4   (a) Coexistence within a crystalline solid of two differently, but "coherently oriented" structures separated by a boundary represented by a straight line. (b) Stacking fault related to a missing atomic plane C in the sequence ABCABC... normally found in a compact FCC cubic structure.

Such a "long range" action, in directions perpendicular to the line, is even more pronounced in the case of dislocations.

Conversely, as shown in Fig. 4.4, the twin boundaries or the stacking faults, are truly two-dimensional since they only disturb the crystal lattice in the immediate neighbourhood of their surface.

Besides dislocations, all the preceding types of defects are involved to a variable degree in the plastic behaviour of a solid. However they do not all have the same importance. The physical study of each type of defect is relatively complex. This is why this chapter will restrict to a single example, the vacancy, which moreover, has a special role in the mechanism of the plastic behaviour. Its mobility controls the motion of dislocations at *high temperatures,* and as a consequence, the plastic behaviour in this temperature range.

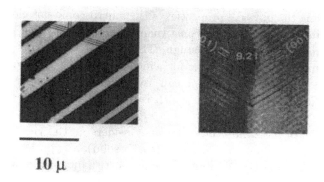

Figure 4.5 (Left) Optical microscope observation of transparent crystal containing a succession of twins separated by parallel boundaries. The dark regions have a common orientation of their lattice. The bright ones have a common orientation distinct from that of the dark regions. (Right) Electronic microscope observation of the boundary between twins. Scale of the striae ~10 Angströms (Thesis of J. Torres Paris 1982).

The study of vacancies will consist, first, in evaluating the energy increase associated to the presence of a vacancy in a solid. The number of vacancies at thermal equilibrium will be deduced from this evaluation. The jump of a vacancy from one crystal site to another will then be examined and the energies involved in such a jump quantified. The collective migration of vacancies in a solid will be determined. The macroscopic effect of this collective motion constitutes the phenomenon of *diffusion*.

4.2 Stability of point-defects in solids

As mentioned in the preceding section, the presence of a point defect in a solid increases its energy. However, it also *increases its entropy*. At non-zero absolute temperature, this entropic contribution favors the stable existence of point defects in the solid.

4.2.1 *Statistical equilibrium*

The thermal equilibrium of a macroscopic system (the set of defects in a solid in the present case) can be described with the help of different *statistical models of its microscopic state*. Fortunately, all these models (e.g.

canonical, microcanonical or grand canonical,...) lead to the same predictions regarding the physical properties of the macroscopic sample, provided that the number of microscopic constituants (atoms, molecules, defects) of the considered sample is large enough. This situation is always realized in the solid samples considered here, for which this number is comparable to the Avogadro number. Hence, one can use indifferently any of these models, and, in practice, use the one which is most convenient for the system considered.

The model of the *canonical ensemble* considers, in the view of describing a given macroscopic "real" system (e.g. a solid), a hypothetical system having a definite number N of microscopic constituants of the same nature as those of the real system. Its total energy can be equal to a large number of possible values labelled U_m, each of these values having the probability $p(U_m)$ of being realized. The *statistical average* $< U >$ of the energy is assumed to be imposed. Its value is considered to be identical to the value U of the energy of the real system which one is able to measure by usual experimental methods performed at the macroscopic level. Besides, one usually deals with systems in which a very large number $W(U_m)$ of distinct configurations C of the set of microscopic constituants have the same total energy U_m. At a temperature $T = (1/k_B\beta)$ the so-called *canonical equilibrium* is defined by a probability distribution $p(C)$ of the configuration C:

$$p(C) = \frac{\exp[-\beta U_m(C)]}{Z_c} \tag{4.1}$$

The probability to measure a given value of the total energy U_m is therefore $p(U_m) = W(U_m).p(C)$. The normalization factor Z_c in the above formula, called the partition function, is determined by:

$$Z_c = \sum_C \exp[-\beta U_m(C)] = \sum_{U_m} W(U_m)\exp[-\beta U_m(C)] \tag{4.2}$$

It is easy to show, on the basis of (4.1), that the value of the energy of the system, at thermal equilibrium is:

$$U =< U >_{eq} = \sum_{U_m} p(U_m).U_m = -\frac{\partial Log Z_c}{\partial \beta} \tag{4.3}$$

and that the entropy of the system is:

$$S = -k_B \sum_{(C)} p(C) Log[p(C)] = k_B Log Z_c + k_B \beta U \tag{4.4}$$

From these two formulas one draws the expression of the free-energy:

$$F_{eq} = U - TS = -\frac{1}{\beta} Log Z_c \tag{4.5}$$

4.2.2 *Concentration of defects at thermal equilibrium*

Assume that the set of defects in a solid constitutes an isolated system possessing a proper total energy. Actually, this energy is determined by the interaction of each defect with its environment in the solid. However, if the concentration of defects is low enough, as to ensure that the energy of each defect is not influenced by the other defects, one can associate an identical energy u_F to each defect. This energy can be defined as the increase of energy of the entire solid when one defect is present. It is shown in section 4.3.1 hereunder, that, for a point defect, this energy has a finite value. Assume also that a volume Ω of the crystalline solid has N *identical* sites in which a defect can be located. For instance in the case of substitutional impurities, N is the number of atoms which are likely to be replaced by an impurity. Likewise, if we consider interstitial atoms, N will be the number of sites in which the "foreign" atom can be inserted.

In this theoretical framework, it is possible to determine the number n_{eq} and the concentration n_{eq}/N of defects at thermal equilibrium. Note, first, that the increase in energy of a solid containing n defects is $U_n = n.u_F$. For a given number n of defects and N possible locations for each defect there are $W(U_n) = C_N^n$ distinct configurations for the identical defects. Each configuration is defined by the location of the sites which are occupied by a defect. The number n being into a one to one correspondence with the value of the total energy, all the $W(U_n)$ configurations have the same energy. Hence the partition function can be written as:

$$Z_c = \sum_{n=0}^{N} C_N^n \exp(-\beta n.u_F) = [1 + \exp(-\beta u_F)]^N \qquad (4.6)$$

from which we deduce:

$$n_{eq} = \frac{U}{u_F} = -\frac{1}{u_F}\frac{\partial Log Z_c}{\partial \beta} = \frac{N e^{-\beta u_F}}{1 + \exp(-\beta u_F)} \approx N \exp(-\beta u_F) \qquad (4.7)$$

The approximation leading to the form of the last term, is justified by the assumption that the concentration (n_{eq}/N) of defects is small. Its consistency can be checked by calculating the value of (n_{eq}/N) determined by the above formula.

Hence, if u_F has a finite value, the atomic concentration (n_{eq}/N) is non-zero, at thermal equilibrium. The exponential functional form $n_{eq} \approx \exp(-u_F/k_B T)$ implies that this concentration increases rapidly with an increase of the temperature. Table 4.1 shows the order of magnitude of the concentration for different characteristic temperatures and defect energies.

Table 4.1 Atomic concentration (n_{eq}/N) of point defects as function of the energy of formation u_F and of the temperature.

$$
\begin{array}{c|c|c|c}
u_F\downarrow \quad T \rightarrow & 300K & 1300K & 1800K \\
\hline
0,5eV & 4.10^{-9} & 10^{-2} & 4.10^{-2} \\
\hline
1eV & 1,7.10^{-17} & 1,4.10^{-4} & 1,6.10^{-3} \\
\hline
2eV & \approx 0 & 2.10^{-8} & 3.10^{-6}
\end{array}
\tag{4.8}
$$

4.3 Formation of vacancies

The absence of an atom on a structural site normally occupied in the ideal crystal structure can be the result of various speculative mechanisms. It can, for instance, be determined by successive *jumps* of atoms. Thus, an atom close to the surface of the crystal sample (Fig. 4.6), can jump into a site of the surface, leaving a vacancy in its initial position. Another atom

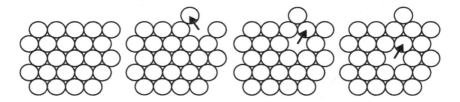

Figure 4.6   Creation of a vacancy inside a crystal by successive atomic jumps.

can then jump into the site of this vacancy, thus displacing the location of the vacancy. Progressively, this mechanism is able to produce a vacancy at an arbitrary atomic site internal to the crystal. Another mechanism, is the result of jumps of an atom into neighbouring interstitial sites. It will be shown to be less probable, at the end of the chapter. One then obtains a so-called "Frenkel defect" (Fig. 4.7) which is composed by a vacancy and of an interstitial impurity located in neighbouring sites. Such a "composite" defect can subsequently dissociate, if the vacancy or the interstitial migrate into more distant sites.

Irradiating a crystal with a beam of particles of suitable energy (electrons, atoms, neutrons) is also likely to produce vacancies by colliding with atoms and displacing them, thus creating Frenkel defects.

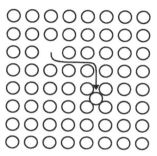

Figure 4.7   Frenkel defect constituted by the displacement of an atom into an interstitial site thus forming a pair vacancy-interstitial.

4.3.1   *Formation energy*

The formation energy u_F of a vacancy in a crystal sample is, by definition, the increase of energy of the sample due to the presence of a vacancy. It is customary to consider that, more precisely, u_F corresponds to the situation in which the displaced atom is evacuated towards the surface of the crystal sample. This increase of energy depends of the nature of the chemical bonds ensuring the cohesion of the crystal.

Before examining the qualitative determination of u_F, in the simple case of a metal, let us evaluate this energy through consideration of a purely mechanical model. Although somewhat artificial, such a model, has the advantage of resembling the more relevant mechanical model of a dislocation, studied in chapter 6. Recent atomistic models of vacancies, based either on classical mechanics or on quantum mechanics, as well as on computer simulations, yield more precise values of u_F.

Description of the elastic model

At the site of a vacancy, there is a complex change of chemical bonding (angles and distances) between the neighbouring atoms. It can have the effect either to increase or to decrease the distances between the atoms surrounding the vacancy. An increase of the interatomic distances will have the effect that the "hole" which has replaced the missing atom occupies a larger volume than the atom itself. A converse effect accompanies a decrease of the interatomic distances. Hence, a *local change of volume* generally occurs at the site of a vacancy.

This remark underlies an "elastic" model aimed at determining the formation energy u_F of a vacancy. In this model, one makes abstraction of the microscopic origin of u_F, and relates its value to the sole fact that the vacancy occupies a different volume than the missing atom. Internal strains and stresses, distributed in the entire solid, result from it, which determine a certain amount of elastic energy. This energy is identified to u_F. The range of the elastic interaction, very large as compared to the interatomic distances, justifies that in this "elastic" model, a solid is assimilated to a continuous medium having no microscopic structure (Fig. 4.8).

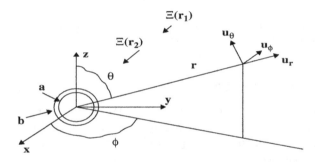

Figure 4.8   Displacement and strain fields induced by a vacancy in a continuous medium.

In this medium, assumed to be infinitely extended, the local change in volume is mimicked by drilling in the continuous matter a spherical cavity having the radius a of the missing atom. Using the deformability of the surrounding matter, a rigid sphere of radius $b \lesssim a$ is inserted in the cavity, and the continuity of the medium restored by "sticking" the surface of the sphere to the internal surface of the cavity. Depending if b is smaller or larger than a, an expansion, or a compression, of the surrounding elastic medium will be induced.[2]

---

[2]More elaborate models have been described in which, for instance, one inserts, instead of a rigid sphere, a deformable one, having the same elastic properties as the surrounding medium. Different values of u_F are then obtained. However, as all these mechanical models do not express the true physical sense of u_F, we only describe here the simplest of these models.

Let us first calculate the displacement field associated to the insertion of the rigid sphere.

Displacement field

The spherical geometry of the preceding model implies that the local displacement of matter be *radial*: $\overrightarrow{\xi(\vec{r})} = \Xi(r)\vec{r}$, the function $\Xi(r)$ being defined only for $r \geqslant b$. On the sphere of radius b, the displacement imposed by the rigidity of the inserted sphere is equal to [3] $(b - a)$.

The displacement $\Xi(r)$ at any point in space must be a solution of the equilibrium Eq. (3.30), compatible with the limit conditions $\Xi = (b - a)$ on the surface of the defect, and $\Xi = 0$ at infinity. A unique solution exists which complies with the additional condition $div\,\overrightarrow{\xi} = 0$, i.e. with no local change of the volume. The extension or contraction in the radial direction is compensated by a contraction/extension in the tangential directions. This solution is:

$$\Xi(r) = \frac{b^2(b-a)}{r^3} = \frac{A}{r^3} \tag{4.9}$$

Indeed, taking into account $div\,\overrightarrow{\xi} = 0$, Eq. (3.30) reduces to $\Delta\,\overrightarrow{\xi} = 0$, or:

$$\frac{d}{dr}\left\{ \frac{1}{r^2}\frac{d}{dr}[r^3\Xi(r)] \right\} = 0 \tag{4.10}$$

It is easy to check directly on expression (4.9) that $div(\Xi.\vec{r}) = 0$. On the other hand, for small values of $(b - a)$, the quantity $4\pi A$ represents the difference of volume between the inserted sphere and the missing atom. $A > 0$ for a local dilatation $(b > a)$ and $A < 0$ for a local contraction $(b < a)$.

Induced strain and stress

From Eq. (4.9) one deduces, in a cartesian frame of reference, the *strain components* at any point $\vec{r}$. These can be written $\epsilon_{ij} = (1/2)(\partial\xi_i/\partial x_j + \partial\xi_j/\partial x_i)$ with $\vec{r} = \{x_1, x_2, x_3\}$. Thus:

$$\epsilon_{ii} = \frac{A}{r^3}[1 - \frac{3x_i^2}{r^2}] \qquad \epsilon_{ij}(i \neq j) = \frac{-3Ax_ix_j}{r^5} \tag{4.11}$$

---

[3] In chapter 5, a similar "continuous model" will be described in the case of dislocations. However in this case, the displacement on the surface of the defect will be a uniform translation instead of being a radial displacement.

From these expressions one can check the invariance of the local volume, around each point external to the core of the defect, in the form $\delta\Omega = \overrightarrow{div\,\xi} = \sum \epsilon_{ii} = 0$. The physical meaning of the strain induced by the vacancy is clearer when a local rectangular frame adapted to the spherical coordinates is used (Fig. 4.8). The following expressions are deduced from (4.11) and from Table 3.2:

$$\epsilon_{rr} = \frac{-2A}{r^3} \;;\; \epsilon_{\theta\theta} = \epsilon_{\phi\phi} = \frac{A}{r^3} \;;\; \epsilon_{r\theta} = \epsilon_{r\phi} = \epsilon_{\theta\phi} = 0 \qquad (4.12)$$

ϵ_{rr} represents the relative change in length in the radial direction. Consistently, a rigid sphere smaller than the missing atom ($A < 0$) produces an extension of the medium along the radius ($\epsilon > 0$), while a sphere larger than the missing atom induces a local contraction along the radius. In this system of coordinates, all *shear components* are equal to zero. Note that the non-zero components decrease as $(1/r^3)$ as a function of the distance to the vacancy.

On the other hand, Eq. (3.26) determines that the stress components are proportional to the corresponding strain components $\sigma_{ij} = 2\mu\epsilon_{ij}$, μ being the Lamé coefficient relative to the shear.

Elastic energy of a vacancy

The elastic energy induced by the vacancy in the entire volume of the solid ($r > b$) is (Eq. 3.5):

$$\frac{u_F}{2\pi} = \int_b^\infty \left\{ \sum \sigma_{ij}\epsilon_{ij} \right\} r^2 dr = 12\mu A^2 \int_b^\infty \frac{dr}{r^4} = \frac{4\mu A^2}{b^3} \qquad (4.13)$$

$$A^2 = \frac{3}{4\pi} \omega_{lac}.b(b-a)^2 \qquad (4.14)$$

The value of μ is generally in the range $(3 \pm 1).10^{10} J/m^3 \approx (0,2 \pm 0,1)eV/\text{Å}^3$ (cf. Table 3.3), and the volume ω_{lac} of a vacancy of the order of ~10Å^3. The volume change associated to a vacancy can be determined experimentally (see for instance exercise n° 3). In certain metals this change corresponds to $|b - a|/b \approx 25\%$. Such a value will determine, on the basis of Eq. (4.12) a strain value of the order of 10^{-3} at a distance $r = 10b$ from the vancancy site, and an energy u_F (cf. Eq. (4.13)) of the order of 0.1 eV. As will be shown in the next paragraph, the actual energy of a vacancy,which can also be determined experimentally (cf. exercise 3), is rather of the order of 1 eV. This significant difference shows the inadequacy of the elastic model.

Energy of a vacancy in a metal

In this paragraph the physical origin of the energy of a vacancy in a metal is qualitatively discussed.

As briefly recalled in section 2.5.1, the binding of a metal is due to its free-electron "gas". For instance in a monoatomic metal (i.e. constituted by a single type of atoms), the structure consists of positive ions, all identical, of electric charge $+Ze$, interacting with the negative electron gas. The electric neutrality is preserved at each point of the metal.

Consider two hypothetical situations for the creation of a vacancy. First assume that an atom migrates towards the surface of the sample with all its valence electrons, thus leaving no electric charge on the site of the missing atom (Fig. 4.9 (a)). Such a circumstance would induce a strong electronic density gradient in the vicinity of the vacancy. (Fig. 4.9 (c)). The quantum mechanical expression of the density is proportional to the square $|\psi|^2$ of the modulus of the electrons wavefunction. Thus, a high density gradient is associated to a high value of the gradient of ψ, hence (through the quantum mechanical correspondance $\overrightarrow{p} \longleftrightarrow (\hbar/i)\overrightarrow{\nabla}\psi$) to a large value of the kinetic energy $p^2/2m$ of the electrons. For instance, a confinement of the gradient on a distance of $(r_0/10)$, determines a kinetic energy of the order of ≈ 30 eV.

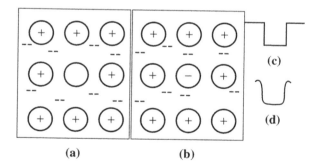

Figure 4.9 Positive ions and uniform electron density in a metal. (a) Model of vacancy assuming the absence of charge at the site of the missing ion. (b) Uniform electron density also filling the vacancy site which thus induces a local excess of charge (-Ze) at this site, the remaining solid being neutral. (c) Abrupt density gradient in model (a). (d) Density redistribution determining a smaller gradient.

The opposite situation in which a uniform electron density is preserved in the entire volume, *including the site of the vacancy,* is not favourable either (Fig. 4.9 (b)). Indeed, the absence of the positive ion has the consequence of determining a *negative charge* $-Ze$ on the site of the vacancy. A repulsion between this localized charge and the surrounding electron gas would then increase significantly the energy of the solid.

The excess energy is minimized by decreasing both the localized charge and the gradient of electron density through a charge distribution of the type represented in Fig. 4.9 (d). For such a distribution, the excess energy is of the order of $4 - 5$ eV for monovalent metals such as gold or copper.

However, this excess energy is partly compensated by the fact that the migration of an atom at the surface of the sample (thus creating the considered vacancy) (Fig. 4.6) *decreases the energy of the sample through an increase of its volume.* Indeed, it can be shown that the electronic energy of an electron gas, in a metal, is:

$$U = \frac{3}{5}\frac{\hbar^2}{2m}N[3\pi^2\frac{N}{\Omega}]^{2/3} = \frac{3}{5}NE_F \qquad (4.15)$$

in which N is the number of free electrons, and Ω the volume of the sample. E_F is (at sufficiently low temperatures) the energy of the occupied electron state with the highest energy. Hence, as Ω increases approximately by one atomic volume ω_{at}, the energy decreases by the amount:

$$\Delta U = \frac{\partial U}{\partial \Omega}\omega_{at} = -\frac{2}{5}E_F \qquad (4.16)$$

For instance in the case of copper, the value of the decrease is $\approx -3eV$.

Finally the excess energy related to the formation of a vacancy is generally in the range 0.5-1,5 eV. For such values, Table 4.1 indicates that, in the vicinity of the melting point, the proportion of vacancies in a solid is of the order of 10^{-3}.

4.3.2 *Random displacement of vacancies, diffusion*

As mentioned in the beginning of the chapter, a possible mechanism leading to an equilibrium distribution of vacancies in a crystalline solid, is the migration of atoms towards the surface of the sample. This mechanism can also be viewed as the migration of vacancies already existing at the surface. The *diffusion of vacancies* is the macroscopic phenomenon consisting in a *variation of the concentration* of vacancies in a region of the solid through their collective migration from one region to another. As will be shown hereunder, the rate of diffusion increases rapidly with an increase of the temperature of the solid. This will derive from the two elements which determine the mechanism of diffusion: i) Elementary jump of a vacancy from a given site towards a neighbouring site. ii) The kinetics of the collective migration of vacancies, associated to a succession of such elementary jumps.

Frequency of jumps

Figure 4.10 shows schematically the displacement of an atom, labelled **A**, to an unoccupied neighbouring site to the right (i.e. a neighbouring vacancy). The initial state (1) and the final state (3) define, an elementary *jump of the vacancy* to the left.

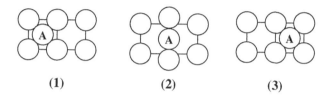

(1) **(2)** **(3)**

Figure 4.10 Hypothetical steps for the jump of an atom between neighbouring sites.

The energy scheme corresponding to this situation is represented in Fig. 4.11. The "ground state" (minimum) energy V_{min} corresponds to the configuration in which the atom **A** is at rest, in either of its two equivalent equilibrium sites. The "excited" energy V_{max} corresponds to the spatial configuration (labelled (2) in Fig. 4.10) in which the atom is in a transient state, on the bottleneck, half way between the two equilibrium sites. An analysis of the conditions in which an atom will jump from one site to the other, through the bottleneck, or alternately, for the vacancy to jump from the right to the left, leads to the result that the average number of jumps per second (i.e. their frequency) effectuated by a vacancy from a site which possesses z neighbouring sites is given by:

$$f_L = \nu z \exp(-\beta E_m) \qquad (4.17)$$

Indeed, to achieve a jump, atom **A** must be in motion towards the left. The required motion is provided by the fact that each atom of the structure *oscillates* around its equilibrium position at a frequency ν. Hence, the atom moves in the direction of the bottleneck ν times per second and has each time an opportunity to jump. In a three-dimensional structure, a vacancy has z neighbouring atoms, and therefore, its probability to jump is multiplied z-fold.

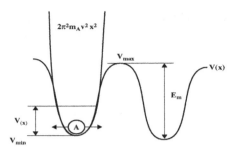

Figure 4.11   Energy barrier E_m opposing the jump of an atom between neighbour sites. The motion of an atom, with frequency ν, about its equilibrium position, will induce, with a non-zero probability, the possibility of overcoming the barrier.

An actual jump will only be possible if the kinetic energy of the oscillating atom is sufficient to overcome the energy barrier $E_m = (V_{\max} - V_{\min})$ corresponding to the bottleneck. At thermal equilibrium, the *average* oscillation energy V_{osc} is not large enough. However, an "excited" oscillation state having the energy E_m is realized with a probability proportional to $\exp(-\beta E_m)$. Atom **A** will then have this probability of being able to jump towards the neighbouring site, thus justifying the form of f_L.

In other terms, it is the thermal motion,[4] in the form of vibration energy, which allows this jump.

The height E_m of the energy barrier is the *migration energy of the vacancy*. The frequency ν is of the order of $10^{13} Hz$. Table 4.2 shows values of f_L for $z = 12$ (e.g. a metal with the FCC structure), for a few selected values of migration energies and of temperatures $T = 1/k\beta$.

Table 4.2 Jump frequency of a vacancy as a function of the migration energy and of the temperature.

$$
\begin{bmatrix}
E_m \downarrow \quad T \rightarrow & 300K & 1300K & 1800K \\
& & & \\
0,2eV & 5.10^{10}s^{-1} & 2.10^{13}s^{-1} & 3.10^{13}s^{-1} \\
0,5eV & 5.10^{5}s^{-1} & 10^{12}s^{-1} & 5.10^{12}s^{-1} \\
1eV & 2.10^{-3}s^{-1} & 1,5.10^{10}s^{-1} & 2.10^{11}s^{-1}
\end{bmatrix} \quad (4.18)
$$

---

[4]The Boltzmann distribution is the statistical expression, at equilibrium, of the physical phenomena, controlled by the thermal motion (e.g. oscillations, collisions), which determine the onset of the thermal equilibrium.

Average free path of the vacancies

Let us place ourselves in a two dimensional crystal in which the atomic sites form a square lattice (Fig. 4.12) and consider the displacement of a vacancy. At each jump, the vacancy has identical probabilities of jumping into any of the four surrounding sites which are at the same distance δ. In this *random walk*, the probability is the same for a displacement $\vec{\delta}_i$ and for the opposite displacement $(-\vec{\delta}_i)$. This symmetry has the consequence that the *statistical average* of the overall displacement of the vacancy through its successive jumps is equal to zero: $< \vec{d}_n > = \sum_1^n < \vec{\delta}_i > = 0$.

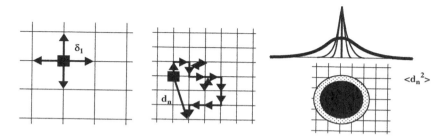

Figure 4.12 Random walk on a square lattice. (Left) Equally probable jumps of a vacancy on its four neighbouring sites. (Center) Relation between the displacement d_n of the vacancy and the number of nearest neighbour jumps, (Right) Progressive spreading, expressing the evolution of $< d_n^2 >$, of the probability to find the vacancy at a given distance of the initial site.

By contrast, the statistical average $< \vec{d}_n^2 >$ of the square of the displacement is non-zero. Indeed, let $\vec{d}_{n-1}$ be the displacement of the vacancy after $(n-1)$ successive jumps in random directions. The displacement $\vec{d}_n$ can be written $\vec{d}_n = \vec{d}_{n-1} + \vec{\delta}_i$ where $\vec{\delta}_i$ is one of the four possible displacements having the $(n-1)$th position as its starting site. The quadratic average of $\vec{d}_n$ is then, assuming all the jumps statistically independent from each other:

$$< \vec{d}_n^2 > = < \vec{d}_{n-1}^2 > + < \vec{\delta}_i^2 > + 2 < \vec{d}_{n-1} > . < \vec{\delta}_i > \qquad (4.19)$$

The last term is zero due to the symmetry of the jumps. Hence, applying the preceding equation from $n = 1$ to $n = n$, one obtains:

$$< \vec{d}_n^2 > = n\delta^2 \qquad (4.20)$$

This result means that although the average position of the vacancy is identical to its initial site, the probability of finding it on another site *spreads* progressively in all directions around the initial site. The radius of the "spreading" is proportional to $\sqrt{n}$ or, alternately, to $\sqrt{t}$ since the number of jumps is proportional to the time which has elapsed. Hence, while the vacancy was confined to the initial site, its presence becomes progressively more uniform in the system. At infinite times, it will have the same probability of being found at any site of the lattice.

Taking into account the frequency f_L of the jumps, the *free path* $l_c = \sqrt{<\vec{d}_f^2>}$, i.e. the radius of "spreading" after one second, is :

$$l_c = \sqrt{<\vec{d}_f^2>} = \delta \sqrt{f_L} \qquad (4.21)$$

Considering the values of f_L indicated in Table 4.2, it appears that, for instance for $E_m \approx 1eV$, the vacancy does not move at room temperature, while at high temperatures it can be found, after a second, at a distance $10^5 \delta \approx 20 - 100$ microns.[5] Note that in this lapse of time, the vacancy will have made 10^{10} jumps of a few Angströms and thus effected a distance of $\approx 20 - 100$ mètres.

Macroscopic diffusion of vacancies

For a solid sample of macroscopic dimensions one considers, within the solid, volumes of matter very large with respect to the volume of an atom (or alternately linear scales very large as compared to interatomic distances). In such volumes, the presence of vacancies is characterized by their local concentration $c(\vec{r}) = n/\delta\Omega$, where n is the number of vacancies in the small volume $\delta\Omega$, surrounding point $\vec{r}$. At thermal equilibrium at the temperature T, this concentration is spatially uniform in the solid, since the probability to find a vacancy is the same at any site of the crystal. In an out-of-equilibrium situation, $c(\vec{r})$ is not necessarily uniform, and, in general, it is time dependent.

If, at at a given time $t = 0$, a non-uniform concentration $c(\vec{r}, t = 0)$ exists in a system, a collective migration of vacancies, a *diffusion,* will take place, consisting, at the microscopic level, in the random displacement of vacancies analyzed in the preceding paragraph. This phenomenon will

---

[5] Such a value is larger than the usual diameter of the grains which constitute metalic samples. hence, in a very short time a vacancy will have a significant probability of having "visited" all the sites of a grain.

generally determine an evolution of the concentration $c(\vec{r}, t)$ as a function of time. Let us establish the equation which governs this evolution.

Consider the example of a FCC cubic monoatomic structure (Fig. 4.13). Examine the diffusion of vacancies in a direction x perpendicular to the dense planes of the structure. In this *one-dimensional* problem, assume that the vacancy concentration is of the form $c(x, t)$. At the atomic scale, the variation along x implies that the successive atomic planes $p-1, p, p+1$ have different vacancy concentrations denoted $c_{p-1}(t), c_p(t), c_{p+1}(t)$ (with values close to each other).

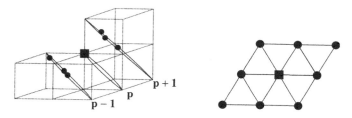

Figure 4.13 Diffusion in the direction normal to the dense atomic planes in a FCC structure. A vacancy, represented by a black square can jump into any of 12 neighbour sites, 6 of which are in the plane p, 3 are the adjacent plane $(p-1)$ and 3 in the plane $(p+1)$.

The time evolution of these planar concentrations is determined by the random jumps of vacancies towards neigbouring sites in other planes. Each vacancy in plane p has twelve neighbouring atoms at distance δ. Six are located in the same plane, while three belong to plane $(p-1)$ and three to plane $(p+1)$. The probability of a jump to any neighbouring site being the same, the probability of a jump towards each of the two adjacent planes is $\frac{1}{4}$. The same probability distribution is valid for vacancies in the planes $p-1$ and $p+1$. Hence, the equation which determines the temporal evolution of the number $n_p(t)$ of vacancies in plane p can be written as:

$$\frac{dn_p}{dt} = [\frac{n_{p+1}}{4} + \frac{n_{p-1}}{4} - 2.\frac{n_p}{4}].f_L \qquad (4.22)$$

where f_L is the frequency of jumps. The volume concentrations $c_p(t)$ being proportional to the n_p, the evolution equation for c_p is:

$$\frac{dc_p}{dt} = [\frac{c_{p+1}}{4} + \frac{c_{p-1}}{4} - 2.\frac{c_p}{4}].f_L \qquad (4.23)$$

From a macroscopic standpoint, the successive atomic planes form a quasi-continuum of planes distant from each other by $dx = \delta\sqrt{2/3}$ (where δ is the distance between neighbouring vacancy sites). The preceding equation can then be expressed as:

$$\frac{\partial c(x_p, t)}{\partial t} = \frac{f}{4}\{[c(x_p + dx, t) - c(x_p, t)] - [c(x_p, t) - c(x_p - dx, t)]\} \quad (4.24)$$

Noting that $\delta = a_c/\sqrt{2}$, where a_c is the size of the conventional cubic unit cell (which was denoted **c** in chapter 2) this equation takes the form:

$$\frac{\partial c(x, t)}{\partial t} = \frac{f_L \delta^2}{6}\frac{\partial^2 c(x, t)}{\partial x^2} = f_L \frac{a_c^2}{12}\frac{\partial^2 c(x, t)}{\partial x^2} \quad (4.25)$$

This result can be generalized to a diffusion in three-dimensions in the form of *Fick equation*:

$$\frac{\partial c(\vec{r}, t)}{\partial t} = \gamma f_L \delta^2 \Delta c(\vec{r}, t) = D_L \Delta c(\vec{r}, t) \quad (4.26)$$

The diffusion coefficient D_L (expressed in $m^2.s^{-1}$) is proportional to the jump frequency f_L and to the square of the distance δ between neighbouring sites. The geometrical factor γ characterizes the specific crystal structure considered. D_L is strongly dependent of the temperature through its proportionality to f_L:

$$D_L = \gamma z \nu \delta^2 \exp(-\beta E_m) \quad (4.27)$$

Note that the same argument which leads to Eq. (4.23), can be used to evaluate the flow J of vacancies in the direction normal to the dense planes of the structure. Let (dn/dt) be the number of vacancies exchanged between plane $(p - 1)$ and plane p across a surface S. One can write:

$$J = \frac{1}{S}\frac{dn}{dt} = \frac{1}{S}[\frac{f_L}{4}(c_{p-1} - c_p)d\Omega] \quad (4.28)$$

in which c_p is the concentration of vacancies per unit volume. Considering the planes as a quasi-continuum, as in Eq. (4.24), with $d\Omega = Sdx$, where dx is the distance between planes, this equation becomes:

$$J = -\frac{f_L \delta^2}{6}\frac{\partial c(x, t)}{\partial x} = -D_L \frac{\partial c(x, t)}{\partial x} \quad (4.29)$$

and, in the general three-dimensional case:

$$\vec{J} = -D_L \vec{\nabla} c(\vec{r}, t) \tag{4.30}$$

in which $\vec{J}$ is the *flow of vacancies* in the considered direction. The latter relationship is the expression of *Fick's law*.

Self-diffusion of atoms

Instead of considering vacancies in an atomic structure, one can examine the diffusion of the atoms themselves. Even in a monoatomic structure, in which all atoms are identical, the jump of an atom from one atomic site to another can be given a meaning. Indeed, certain atoms of the structure can be "marked" by transforming them into one of their isotopes through irradiation by a flow of specific particles. This will not change their electronic properties, hence their bonding with neighbouring atoms. The jump of a "marked" atom from one site to another can then be studied.

Although the jump of a vacancy towards a neighbouring site occupied by an atom consists in the jump of an atom into the vacancy site, the frequency f_A of atomic jumps is different from that f_L of vacancies. Indeed, an atomic jump is only possible if a vacancy exists in its neighbourhood. The probability of this event, for a crystal possessing N atomic sites is $(z n_{L/N})$, in which n_L is the number of vacancies in the crystal at thermal equilibrium, and z is the number of atomic sites around an atom. On the other hand, by contrast to the vacancy which can jump to any of the surrounding atomic sites, a single possibility exists for an atom (neglecting the probability of having two vacancies next to the atom). Thus, combining Eqs. (4.7) and (4.8), with $u = u_F$ (Eq. (4.13)):

$$f_A = \frac{f_L}{z} \cdot \frac{z n_L}{N} = \nu z \exp[-\beta(E_m + u_F)] = f_L.c \tag{4.31}$$

Since the other arguments developped for vacancies, concerning their random walk and the collective diffusion, can be reproduced identically for atoms, Eqs. (4.26) and (4.30) are valid for the diffusion of atoms, using a *self-diffusion atomic coefficient*:

$$D_A = \gamma f_A \delta^2 = \gamma z \nu \delta^2 \exp[-\beta(E_m + u_F)] \tag{4.32}$$

Table 4.3 hereunder compares the values of the diffusion coefficients for vacancies and for atoms at different temperatures for $E_m \approx 0.8$ eV, $u_F \approx 1,1$ eV, $\gamma = (1/6)$, $z = 12$, $\nu = 10^{13}$ Hz, $\delta = 3.10^{-10}m$.

Table 4.3

$$
\begin{bmatrix}
\begin{array}{c|c|c|c}
T \rightarrow & 300K & 1000K & 1300K \\
\hline
D_L(cm^2.s^{-1}) \rightarrow & 6.10^{-21} & 5.10^{-8} & 10^{-6} \\
D_A(cm^2.s^{-1}) \rightarrow & 2.10^{-34} & 5.10^{-12} & 8.10^{-10}
\end{array}
\end{bmatrix}
$$

Other types of point defects

The preceding results can be extended to other types of point defects, by taking into account, each time, the specific values of the characteristic energies. For instance, atoms located on intersticial sites can jump towards neighbouring interstitial sites. Such jumps will allow their displacement through the solid in order to reach a uniform concentration at equilibrium. However, the formation energy of an interstitial atom is significantly different from that of a vacancy. Consider, for instance, a monoatomic metal with the FCC structure. As mentioned in chapter 2 (section 2.4.2), the octahedral interstitial sites can contain atoms with a maximum radius $\rho = 0,414\mathbf{r}$, $\mathbf{r}$ being the radius of a constituting atom of the metal. Using the elastic model developed in section 4.3.1, we can evaluate the elastic energy u_F for an atom displaced into an intertitial site. The ratio $|(b - a)/b|$ appearing in Eq. (4.13) is $|(\mathbf{r} - \rho)/\mathbf{r}| \approx 0,6$. From this value, one determines $u_F \approx 2 - 4$ eV, an order of magnitude larger than the elastic energy detrmined for a vacancy. For such a value the defect concentration at equilibrium is negligible, even at temperatures as high as the melting temperature of the solid (e.g. $c \approx 5.10^{-12}$ at $T = 1800$ K).

Chapter 5

The geometry of dislocations

Main concepts: Straight dislocations of the edge and screw types. Crystallographic and Volterra descriptions. Formation by gliding or by addition or substraction of matter. Characterization of the sign and "strength" of a dislocation by a Burgers vector $\vec{b}$. Dislocation loops. (b/r) law of decrease with distance of the deformation induced by a straight dislocation. Proportionality to b^2 of the energy density associated to a dislocation. Electron microscopical method of observation of dislocations.

5.1 Introduction

The straight "edge" dislocation briefly defined in chapter 4 is reproduced again in Fig 5.1. The terminology used for this defect comes from its description as a straight line limiting a half atomic plane within the crystalline solid, similar to the "edge" of a knife inserted into the solid.

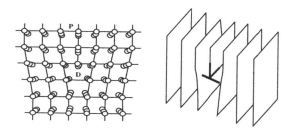

Figure 5.1 Edge-dislocation obtained by inserting an additional half atomic plane P. (Right) Simplified representation in which the symbol ⊥ represents an edge dislocation having the additional half-plane above the symbol.

Pursuing this analogy, we can consider, on a purely hypothetical basis, that this defect has been formed by separating two adjacent half planes of the ideal crystal and inserting an additional half plane (Fig. 5.2).

The region labelled D, of diameter equal to a few interatomic distances, surrounding closely the lower edge of the additional half plane, is called the *core* of the dislocation. In this region, the configuration of bonds between neighbouring atoms is strongly modified with respect to the situation of the ideal structure. The representation of the core on Fig. 5.1 is schematic. The actual configuration, which will be discussed in chapter 7, is not easily accessible experimentally. Theoretical or numerical-simulation models of its structure are subjects of current research.

Outside the core, the configuration of bonds, although deformed, has a close similarity with the ideal crystal. The length or angle changes have the same magnitude as the changes imposed by a moderate external mechanical stress imposed to the solid. This moderately deformed region is labelled the "good crystal" region, by contrast to the strongly perturbed region D of the core.

The straight edge dislocation is a particular example of a class of linear defects, called *dislocations*, which do not all have the shape of straight lines. The various members of this family of defects differ by the crystal modification they induce. However, they all share a common definition and a common theory. The present chapter is devoted to their general geometrical characteristics, relying, as a first step, on the exemple of the straight edge dislocation.

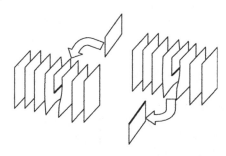

Figure 5.2   (Left) Edge-dislocation formed by inserting a half-plane in the upper part of the crystal. (Right) The same dislocation obtained by withdrawing a half-plane from the lower part.

5.2 Straight edge dislocation

5.2.1 *Hypothetical procedures of formation*

Addition or substraction of a half atomic plane

The straight edge dislocation has been described hereabove as resulting from the presence of an additional half atomic plane (represented in Fig. 5.1 in the upper part of the crystal). The speculative procedure of forming such a defect would be to cut the crystal above line D, separating along this cut the two adjacent half atomic planes, and inserting an additional half atomic plane, and, finally, restoring the chemical bonds between atoms on either sides of the cut. Obviously, the *same defect* can be associated to the absence of a half plane in the lower side of the crystal in Fig. 5.2. The corresponding mechanism of formation would then be to withdraw a half atomic plane, after cutting the bonds with the two adjacent half planes, and subsequent restoring of the bonds between the two remaining half planes facing eachother.

Thus a given dislocation does not result from a unique mechanism of formation. Other hypothetical modes of formation of the same dislocation can be considered, deriving from a different procedure.

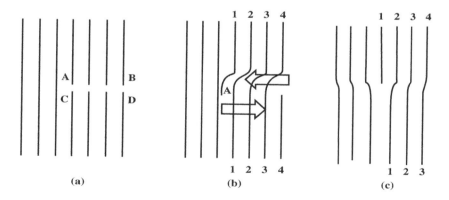

Figure 5.3   Formation of an edge dislocation by the following successive operations: (a) Horizontal cut of the bonds between atoms in the right part of the crystal; (b) Uniform slipping of AB to the left; (c) Restoration of the bonds between atoms facing each other, and relaxation to a symmetric configuration. Thick arrows represent the shear stress needed to achieve this procedure.

Formation by partial slipping

A dislocation can be produced by the following sequence of operations (Fig.5.3) :

a) Cut the crystal along the horizontal half-plane separating the upper and lower halves of the crystal, at the right of the AC line perpendicular to the figure. One thus forms two edges whose projections are AB and CD (Fig. 5.3 (a)).

b) Translate *uniformly* the upper half-plane AB,[1] to the left by *one* interatomic distance. The bonding of the atoms in this plane with the upper part of the crystal will induce a displacement of this part to the left. Such an hypothetical operation can, in principle, be concretely realized by application of a shear stress of sufficient magnitude. However, obviously, this slipping (or gliding) cannot be totally uniform: near point A it will be stopped by the obstacle of the half-crystal situated at the left of A, in which no cut has been effected. As a consequence, in the vicinity of A, there will be a local contraction of the distances between the vertical atomic planes. This "compression" will occur in a region a few interatomic distances in width. This region is to become the core of the dislocation (Fig. 5.3 (b)).

c) Eliminate the cut by restoring the bonds between atoms facing each other, *after slipping*, along AB and CD. However, the vertical half-plane above A has no atom facing it along CD. Hence, the atomic configuration in this region is strongly perturbed with respect to the ideal crystal structure.

d) Suppress the external shear stress which has permitted the slipping. The atomic configuration "relaxes", i.e. takes a new configuration which is symmetric with respect to the vertical plane at A, as represented in Fig. 5.3 (c)).

Clearly, this final configuration is identical to the one obtained, in Fig. 5.1 by addition or substraction of a half atomic plane.

Note that the same dislocation can be "fabricated" by having CD slip to the right by one interatomic distance. The interplane distance near C would then be dilatated. Still another procedure would be to realize a horizontal cut in the crystal, *on the left of the dislocation* and induce a slipping of the upper (lower) part of the crystal to the right (left) with a subsequent contraction (expansion) of the distance between vertical planes in the region of the core.

---

[1]One could, equivalently translate, to the left, the entire quarter crystal situated in the right upper part of the figure. The cut and subsequent restoring of the bonds after slipping being confined to AB, the "elastic relaxation" of the crystal would lead to the same final result.

Whatever the chosen procedure, the dislocation has the same final appearance, after restoration of the chemical bonds and "relaxation" of the crystal. It corresponds to the insertion of a single additional half atomic plane, or to the horizontal slipping of a portion of the crystal by *one interatomic distance*.

Each of these hypothetical procedures has a concrete relevance. For instance, as already mentioned, the formation by slipping can be, in principle, induced by application of a suitably oriented shear stress [2] (Fig. 5.3 (b)).

Amplitude of the slipping and primitive translations

The uniform translation of part of the crystal (except as mentioned in the vicinity of A) determines the formation of a dislocation *only if its magnitude is equal to the distance between parallel atomic planes*. Indeed, this condition ensures the restoration of an approximately "normal" structure everywhere except in the core of the dislocation. Each atom situated along AB is substituted by the neighbouring, identical, atom. If the slipping is not equal to a multiple of the interplane distance, the configuration of the chemical bonds would be strongly modified on the entire half plane projected on AB. The defective region of the crystal would not be reduced to D, and the defect thus generated would not be the linear defect which is being defined here.[3]

Note that, since the slipping distance separates equivalent planes of the ideal structure, this distance is also a *period of the crystal lattice* in the direction of AB, i.e. a combination with integer coefficients of its three primitive translations.

Figure 5.4 shows a slipping by *one* interplane distance. A dislocation of a different nature would be obtained if the amplitude of the slipping was equal to a multiple of the interplane distance, or if several half-atomic planes were inserted (Fig. 5.4). The "core" of this dislocation would be more extended, although its size would still be a few interatomic distances. Clearly, the deformation induced in the surrounding "good crystal" would be larger and would thus determine a larger amount of elastic energy. The

---

[2]It will be shown in chapter 8 that the magnitude of such a stress, required to insert a first dislocation in a crystal is of the order of 10^2-10^3kg/mm^2, thus two or three orders of magnitude higher than the yield strength usually measured. It will also be shown that the procedure involving the insertion of matter (the additional half plane) implies the mechanism of atomic diffusion discussed in the preceding chapter.

[3]Such a planar defect is unlikely to occur owing to the large increase of energy associated to it. However, it will be shown in chapter 7 that certain dislocations include a small portion of planar defect of width a few interatomic distances.

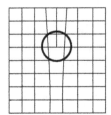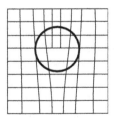

Figure 5.4   Schematic representations of dislocations obtained by insertion of one or two half atomic planes. It will be shown in section 6.5.1 that the dislocation on the right will tend to "decompose" into two dislocations repelling each other and moving far from one another.

occurence of the simplest dislocation, associated to a slipping by a *single* interplane distance will thus be energetically favoured.

General definition of a dislocation

Any type of dislocation shares a common definition with the preceding example of the straight edge dislocation:

A dislocation is a line of discontinuity which separates a portion of a crystal, which has undergone a uniform translation, equal to a lattice-translation of the crystal, from the remaining volume of the crystal assumed to be undisplaced (Fig. 5.3).

In the subsequent paragraphs, it will be shown that each type of dislocation can always be obtained by several methods among which, the slipping of a portion of crystal along a surface of contact, called *the gliding surface of the dislocation* (in the case of the straight edge dislocation of Fig. 5.3, this surface is projected on AB). In the general case, this surface is a cylinder parallel to the direction of the uniform translation imposed, and whose director is the dislocation line (in general not a straight line).

5.2.2   Burgers circuit and Burgers vector

As pointed out, the "good crystal" region external to the core of a dislocation, is similar to a slightly deformed ideal crystal. However, this statement is only true from the standpoint of the *local atomic configuration*. The *global* topology of a crystal containing a dislocation is qualitatively different from that of deformed ideal crystals.

Indeed, consider in such a crystal a *closed loop* surrounding the dislocation, entirely situated within the "good crystal region", and formed by lattice translations joining successive nodes of the lattice (Fig. 5.5). Denote each translation $\overrightarrow{B_i B_{i+1}}$, with B_0 the origin of the closed loop, and B_n the final node which coincides with B_0. Imagine now an identical crystal (having the same atomic structure), *without dislocation*. Consider any node C_0 of its lattice. Owing to the fact that the local configuration of the "good crystal" region of the defected crystal, and that of the ideal one, are similar, it is possible to establish a correspondance between the succession of nodes B_i and a succession of nodes C_i in the ideal crystal such as $\overrightarrow{B_i B_{i+1}} \approx \overrightarrow{C_i C_{i+1}}$, the slight difference between the two vectors being due to the deformation of the defected lattice.

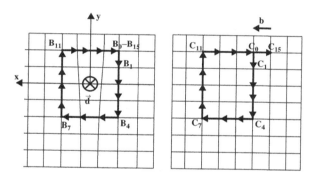

Figure 5.5   Closed loop $B_0 B_{15}$ surrounding an edge dislocation (left) and corresponding loop $C_0 C_{15}$ drawn in the ideal crystal. The latter loop has a closure defect of amplitude b. The vector $\overrightarrow{d}$ orienting the dislocation line is towards the back of the figure and the loop is oriented in the "direct" sense, in agreement with the convention stated in section 5.2.2. The dislocation is assumed to be formed by a slip of the upper part of the crystal to the left.

It appears that the final node C_n does not coincide with C_0: The loop in the ideal crystal *does not close*. There is a closure defect $C_n C_0$. The length and direction of this segment *are the same as those of the uniform translation, equal to a lattice translation, which determines the formation of the dislocation by slipping.* An identical defect of closure would be obtained for any other closed loop, provided that it surrounds the dislocation.[4] Hence, even if one considers a loop passing by nodes situated at a very large dis-

---

[4]The loop can be contained in a plane or be composed of a succession of translations forming a three-dimensional circuit.

tance of the core of the dislocation, in a region unsignificantly deformed locally with respect to the ideal lattice, the defect of closure $C_0 C_n$ will reveal the existence of a dislocation in the crystal, and also its characteristic feature which is the amplitude of the slipping by which it has been formed.[5]

Burgers circuit

Define an orientation of the dislocation line by means of a unit vector $\vec{d}$. From this orientation one can derive an orientation of the closed loops $B_0 B_n$ surrounding the dislocation. Thus, $\vec{d}$ being considered as the z-axis, the loop will be oriented, for instance, according to Maxwell's screw rule. The *oriented loop* $B_0 B_n$, formed by a succession of Bravais lattice translations, is called a *Burgers circuit*. Through the correspondence with $B_0 B_n$, one can likewise orient the line $C_0 C_n$. Given this orientation, the *Burgers vector* $\vec{b}$ of the dislocation is then defined as:

$$\vec{b} = \overrightarrow{C_n C_0} \tag{5.1}$$

$\vec{b}$ is the vector which completes the closure of the open circuit $C_0 C_n$ in the ideal crystal. Since the crystal has an atomic configuration which is invariant parallel to the dislocation line, the same Burgers vector $\vec{b}$ would be obtained if the Burgers circuit is translated parallel to this line. Such an invariance is consistent with the fact that $\vec{b}$ is a measure of the magnitude of the discontinuous translation of part of the crystal, itself constant across the dislocation line.

Note that in the case of the edge dislocation, discussed in this section, the Burgers vector is perpendicular to the dislocation line. This can be expressed by the condition $\vec{b} \cdot \vec{d} = 0$.

Sign of the Burgers vector of an edge dislocation

The modulus of $\vec{b}$, and its direction, do not depend of the orientations chosen for the dislocation line or the Burgers circuit. But its sign (the sense of the vector) depends from these choices. No physical argument imposes this sign, which has therefore to be fixed by a convention. Referring to

---

[5] In another presentation of the Burgers loop, a different procedure is adopted. One then considers a closed loop $C_0 C_n$ in the ideal crystal, and the corresponding path $B_0 B_n$ in the defective crystal containing a dislocation. The possible closure defect will then occur in the dislocated crystal. However, this procedure has two drawbacks. First, this path being open, the fact that it surrounds the dislocation is not defined unambiguously. Besides, the defect of closure $B_0 B_n$, belonging to a deformed lattice is not strictly equal to a translation of the ideal lattice.

Fig. 5.5, the choice of a sign is made as following. An orientation $\vec{d}$ of the dislocation line being chosen, denote $\vec{y}$ the unit vector perpendicular to the glide plane and pointing towards the compressed region of the crystal (i.e. towards the additional half-plane). The sign of $\vec{b}$ is such as to have $(\vec{b}, \vec{y}, \vec{d})$ as a direct frame.[6]

Physical meaning of the Burgers vector

As already stated, in the "good crystal" region, the atomic structure is slightly deformed with respect to the ideal structure. The deformation consists of small variations of the distances between adjacent nodes of the Bravais lattice. Far from the dislocation core, these variations are very small. However their algebraic sum, cumulated over the length of any Burgers circuit surrounding the dislocation, i.e. the quantity $|C_0 C_n - B_0 B_n|$, is precisely equal to the modulus b of the Burgers vector. In other terms b is the measure of the variation of length of a Burgers circuit as compared to its length in the ideal crystal.

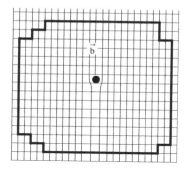

Figure 5.6   Burgers loop drawn at a large distance from the dislocation core, illustrating the (b/r) decrease of the deformation induced by a dislocation.

Since the *absolute variation* b of this length is independent of the Burgers circuit considered, its *relative variation* will decrease when the size of the Burgers circuit increases. This remark provides us with a precise indication on the law governing the decrease of the strain induced by a dislocation with the distance D to the dislocation. Indeed, consider a Burgers circuit,

---

[6]This choice of sign is consistent with the expression which will be given of the important equation due to Peach and Koehler, and which will be introduced in chapter 6. An opposite choice would change a sign in this equation.

approximately circular (Fig. 5.6), centered on the dislocation in a plane perpendicular to its direction, and whose radius is r. The length of this circuit is $\tilde{}\,2\pi r$. Its defect of closure is b, equal to the variation of length of the circuit as compared to the ideal crystal. The average relative change of the distance between adjacent nodes is therefore $\tilde{}\,(b/2\pi r)$. This ratio is equal to the local strain ϵ. Hence, this strain decreases with the distance to D as $\propto (b/r)$. On the other hand, the elastic energy density u generated by the presence of the dislocation, which is proportional to the square of the local strain, and to the elastic stiffness C is proportional to Cb^2.[7]

These variation laws for ϵ and for u will be made more precise in chapter 6.

5.2.3   *Edge dislocation loops*

Rectangular loop

Assume that instead of inserting in a crystal an additional half-plane, one inserts a rectangular portion of plane ABCD (Fig. 5.7 (left)). In the vicinity of each side of this rectangle, the atomic configuration will be that of an edge dislocation since it corresponds to the limit of an additional atomic plane. Hence, the rectangle ABCD is composed of four straight edge-dislocations, each one being a segment of line instead of being an infinite line as in the one described in the preceding section. An edge dislocation loop of rectangular form has thus been generated.

As in the case of the dislocation line, several hypothetical procedures of formation can be considered. In particular, this loop can be formed by a relative slipping of a portion of the crystal. Indeed, one can, for instance (Fig. 5.7 center), maintain fixed the region of the crystal situated at the left of ABCD. One then translates uniformly to the left, parallel to αA, the surface of the cylinder of basis ABCD, the amplitude of this translation being $\overrightarrow{b}$. In practice, this hypothetical operation can be concretely achieved by pushing leftwards the matter internal to the prism ABCDα. The line ABCD is a line of discontinuity of the displacement (amplitude of translation $\overrightarrow{b}$ to the right of this line, zero amplitude to the left).

Note that αA or βB are not dislocations despite the fact that a gliding of the internal volume of the prism $ABCD\alpha\beta$ has been imposed. Indeed,

---

[7]C represents a non-specified elastic stiffness. It will be shown in chapter 6 that for a dislocation the relevant stiffness coefficient is the Lamé coefficient μ.

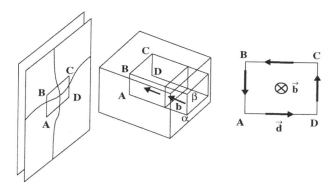

Figure 5.7 (Left) Formation of a rectangular loop of edge dislocations by insertion of a rectangular portion ABCD of lattice plane in a crystal. (Center) Same dislocation formed by slipping of a cylindrical portion of a crystal. (Right) Orientation of the dislocation and of the Burgers vector in accordance with the convention defined in section 5.2.2.

these are not singular lines because, for instance, the crystallographic planes perpendicular to αA are continuous on either sides of this line.[8]

A Burgers circuit can be plotted surrounding any of the edges of the rectangle (e.g. AB) (Fig. 5.7 (right)), and situated, for instance in a plane parallel to αA. Define unit vectors $\vec{d}$, on the four segments of the rectangle, oriented in continuity with each other. Adopting, on each segment the same convention as in the preceding paragraph, for the $\vec{y}$ axis, i.e. a vector pointing towards the compressed region of the crystal, that is, towards the internal surface of the rectangle, and ($\vec{b}$, $\vec{y}$, $\vec{d}$) forming a direct frame, it is easy to check that the sign of the Burgers vector is the same on all four edges of the rectangle. The magnitude of $\vec{b}$ being equal to the uniform translation imposed to form the dislocation-loop, it appears that $\vec{b}$ characterizes the entire rectangular loop. As in the case of the straight edge-dislocation, $\vec{b}$ is perpendicular to each edge of the rectangle: $\vec{b} \cdot \vec{d} = 0$.

Dislocation-loop of arbitrary shape

All the conclusions of the preceding paragraph remain valid if one inserts in the crystal a portion of atomic plane encircled by a planar curve of

---

[8]αA and βB would become lines of discontinuity if one induced the slip of the portions of planes $\alpha A\beta B$ while maintaining the other faces of the prism fixed. It will be shown in section 5.3 that in this case AB would still be an edge-dislocation while αA and βB would be so-called screw-dislocations.

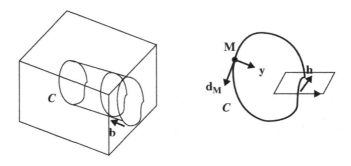

Figure 5.8   Planar edge-dislocation loop.

arbitrary shape C, or, if, equivalently, if one "stamps" the crystal, parallel to $\vec{b}$ with help of a stamp of arbitrary shape (Fig. 5.8), thus inducing a slip of a cylinder of the crystal the base of which is a curve of arbitrary shape. One then obtains a line of discontinuity having this shape. Let $\vec{d}_M$ be a unit vector on the tangent at point M of the dislocation. The vectors $\vec{d}_M$ being oriented in continuity with each other at the successive points of the dislocation, and the compressed region of the crystal being in the internal region encircled by the dislocation, the usual rule determines a unique Burgers vector complying with the condition $\vec{b} \cdot \vec{d}_M = 0$ at any point of the dislocation.

Such a dislocation loop is the most general shape of edge-dislocation.

5.3   Other types of dislocations

5.3.1   *Screw dislocation*

Formation by slipping

A *screw dislocation* is, similarly to the edge dislocation, a line of discontinuity, formed by a slipping of a portion of a crystal, the amplitude of this slipping being a uniform translation $\vec{b}$. However, in this case, the slipping is *parallel* to the line of discontinuity. This implies that a screw dislocation is always a straight line parallel to $\vec{b}$. By contrast, an edge dislocation being a line perpendicular to the uniform slipping $\vec{b}$, its shape can be any line contained in a plane perpendicular to $\vec{b}$. However, it will be shown in the next section that a curve of any shape, not necessarily planar, can be *locally* of the screw type or of the edge-type.

Figure 5.9 illustrates the manner, similar to that of an edge dislocation, by which a screw dislocation can be hypothetically formed.

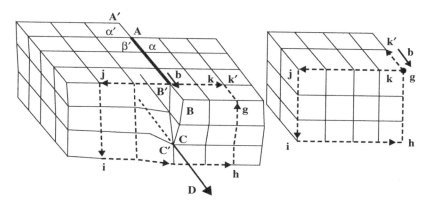

Figure 5.9   Formation of a screw dislocation by slipping parallel to the horizontal dislocation line CD. (Right) Closure defect of the "image" in an ideal crystal, of the Burgers loop kjihgk drawn around the dislocation.

a) Separate the ABC-surface (right) and the A'B'C'-surface (left) of the crystal along the vertical half-plane above the horizontal line D.

b) Impose, by use of external forces, a *forward* slipping of ABC by a distance $\overrightarrow{b}$. This slipping cannot be uniform on the vicinity of D since it vanishes in the fixed part of the crystal situated below D. The atomic planes are therefore deformed in this region.

c) Restore the chemical bonds, such as $A\alpha$ or $B\beta$, between atoms facing eachother and suppress any external force. The atomic configuration "relaxes" towards a new equilibrium configuration which is symmetric with respect to the vertical plane projecting on CC'.

As in the case of the edge-dislocation, a *dislocation-core* can be distinguished as the cylindrical region surrounding D in which the configuration of the bonds is strongly perturbed. The remaining volume of the crystal, the *good crystal region*, although slightly deformed, is topologically similar to the ideal crystal.

A remarkable feature of a crystal containing a screw dislocation is the fact that all the atomic planes perpendicular to the dislocation line are "interconnected" and form, together, a *helicoidal path (e.g. ghijk)* which leads continuously from the bottom of the crystal to its top, through the successive planes. Indeed, *above* the dislocation, the half planes perpendicular to the dislocation such as gB, which were facing the half planes such as jB, prior to the formation of the dislocation, now belong to distinct adjacent planes. By contrast, *below* the dislocation, the half planes containing respectively g and j are still linked to eachother, in the same slightly deformed plane, for instance along ih. In the drawn example, the helicoidal path is a *right handed helix*: travelling in the direction of the vector $\overrightarrow{d}$, one has to "screw" *rightwards* along the path. A left-handed helix would have been obtained if the slipping $\overrightarrow{b}$ had been inverted.

Burgers vector

As in the case of the edge-dislocation, consider a *closed loop kjihgk'* surrounding the dislocation, entirely situated within the "good crystal region", and formed by lattice translations joining successive nodes of the lattice. Orient this loop according to the Maxwell's "screw rule", with the unit vector $\overrightarrow{d}$ of the dislocation line as the z-axis. Note that, for a screw-dislocation, such a circuit cannot be drawn in a plane.

Consider now, in the ideal crystal, starting from an arbitrary lattice node, the homologous sequence of lattice translations corresponding to *kjihgk'*. Again, the final node k' does not coincide with k: The loop in the ideal crystal *does not close*. The closure defect $\overrightarrow{b}$ has the same magnitude as the translation-slip which determines the formation of the dislocation.

For a right-handed screw-dislocation $\overrightarrow{d}$ has necessarily the same sign as $\overrightarrow{b}$. Following the loop in the direction defined by Maxwell's rule consists in following *upwards* (towards the end-point of $\overrightarrow{d}$) the helicoidal path of interconnected lattice planes (Fig. 5.9). In the case of a left-handed dislocation, one goes downwards with respect to the direction of $\overrightarrow{d}$, the Burgers vector $\overrightarrow{b}$ being opposite in direction to $\overrightarrow{d}$. This vector has the same physical meaning as in an edge-dislocation. It is a measure of the cumulated deformations of the lattice translations along a Burgers circuit. Its modulus is equal to the *pitch* of the helix formed by the interconnected path of lattice planes perpendicular to the dislocation.

5.3.2 *Mixed dislocation-loops*

Consider in Fig. 5.10 the rectangular portion ABCD of crystallographic plane internal to a crystal. Realize a cut along this surface, in order to determine two "lips" ABCD (above the surface) and A'B'C'D' (below).

Induce an (almost) uniform slipping of ABCD, parallel to BC, by one interatomic distance b, maintaining fixed the remaining volume of the crystal. Again, the uniformity of the translation is not respected in the vicinity of the four edges of the rectangle. These four segments are lines of discontinuity of the displacement, hence they are dislocations. After restoring the bonds between atoms facing each other on the surface of the rectangle, and suppressing the forces which have permitted the relative displacement of matter, we obtain four segments of dislocations which distort the crystal in their vicinity.

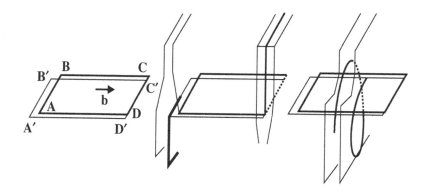

Figure 5.10 Formation of a mixed-type dislocation loop by slipping of the rectangle ABCD parallel to $\vec{b}$. AB and CD are edge-type segments while AD and BC are screw-type segments. (Center) Additional half-planes associated to the edge segments. (Right) Interconnection of lattice planes associated to the screw-type segments.

Along AB and CD the direction $\vec{b}$ of the slipping is normal to the dislocation lines, as standard for an edge-dislocation Consistently, the gliding induces the occurence of an additional half-plane above CD and below A'B'. These two segments of lines are portions of edge-dislocations.

By contrast BC and AD are portions of screw-dislocation lines. Indeed, $\vec{b}$ is parallel to these segments.The slipping shifts to the right, by a translation b, the planes perpendicular to AD situated behind this segment, while it does not modify their position in front of AD. One therefore induces an interconnection between adjacent planes through circuits surrounding AD (cf. Fig. 5.10). A similar situation is established for BC. Note that AD is a left-handed screw-dislocation, while BC is a right-handed one.

ABCD is a rectangular dislocation loop involving two segments which are edge-dislocations and two segments which are screw-dislocations. If AB is oriented by a unit vector $\vec{d}$, pointing towards B, and the other segments are oriented in continuity, we observe that on each segment, the relative orientation of $\vec{b}$ and $\vec{d}$ is in conformity with the conventions underlined above for edge- and screw- dislocations.

If instead of a rectangle we consider a planar closed-curve C of any shape, the same cut and slipping procedure involving a translation $\vec{b}$ parallel to the plane containing the curve, will generate a dislocation loop C. The portions of the loop (possibly infinitesimal) perpendicular to $\vec{b}$ will be of the edge-type, while the portions parallel to $\vec{b}$ will be of the screw-type. The other portions of the loop C are dislocations of *mixed-type*, for which $\vec{b}$ is neither parallel nor perpendicular to the direction of the local segment.

5.3.3   *General properties of the Burgers vector*

Let us summarize the properties of the Burgers vector (Fig. 5.11).

a) As emphasized, its sign depends of the orientation chosen for the dislocation line. Inverting the unit vector $\vec{d}$ tangent to this line, will also invert the sign of the Burgers vector $\vec{b}$. For a given orientation of $\vec{d}$, its sign is *physically* defined for a screw dislocation: $\vec{b} \cdot \vec{d} > 0$ for a right-handed screw-dislocation and $\vec{b} \cdot \vec{d} < 0$ for a left-handed one. By contrast, for an edge-dislocation, its sign depends of a convention, the one adopted here being that $(\vec{b}, \vec{y}, \vec{d})$ forms a direct frame.

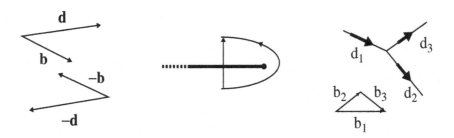

Figure 5.11   Properties of the Burgers vector. (Left) Simultaneous inversion of the signs of $\vec{b}$ and $\vec{d}$ (Center) Impossibility of terminating a dislocation line within a perfect crystal. (Right) Composition rule of the Burgers vectors at the junction of three dislocations.

b) The magnitude of $\overrightarrow{b}$ is constant (uniform) along the dislocation line.[9] This property derives from the fact that b is a measure of the magnitude of the uniform slipping of a portion of the crystal, which has determined the existence of the dislocation. The latter argument also has the consequence that in a crystal which has no other defect than the considered dislocation, the dislocation-line cannot end inside the crystal. This line can either form a loop, or end on the external surface of the crystal. Indeed, consider a Burgers circuit surrounding the dislocation line. By translating it along the dislocation line into the region where the line has ended, this circuit will no longer surround the dislocation. Hence, depending of its position it will either define a non-zero Burgers vector, or a zero Burgers vector. This is in contradiction with the fact that the Burgers vector whose value is associated to the magnitude of a discontinuity does not depend of the position of the Burgers circuit.

Hence, if a dislocation is an open line it can only end on other defects(which will account for the discontinuous variation of $\overrightarrow{b}$): the already mentioned external surfaces of the sample, internal precipitates, grain boundaries, etc... It can also end on one or several other dislocations. The segments of dislocations constituting the rectangular loops considered above, are illustrations of this situation.

If the dislocation lines which meet at a node are oriented in continuity with eachother (Fig. 5.11), one has necessarily:

$$\overrightarrow{b}_1 = \overrightarrow{b}_2 + \overrightarrow{b}_3 \tag{5.2}$$

Since the $\overrightarrow{b}_i$ are lattice translations of the same crystal lattice, the above equality imposes specific restrictions to the types of nodes which can be formed between dislocations (see exercise 9).

5.4 Volterra process of formation

As shown in section 5.2.2, the magnitude of the deformation induced by a dislocation in the good crystal (outside the core) decreases with the distance r to the dislocation as $\epsilon \approx (b/r)$. A significant deformation of the lattice $\epsilon \approx 10^{-3}$ is thus obtained even at distances as large as $r > 100b \approx 500 - 1000\text{Å}$. The mechanical effects of a dislocation will therefore be felt at distances very large as compared to the interatomic distances. At such distances, the deformation will experience small variations from one unit cell to the next

---

[9]This uniformity is also valid for the sign of $\overrightarrow{b}$ if the dislocation is uniformly oriented.

one. If one is not interested in the phenomena occuring within the core of the dislocation, it is thus relevant to study dislocations independently from the atomic structure, thus assimilating the crystal sample to a continuous medium. It is such an approach which underlies the so-called *Volterra process* of formation of a dislocation. Such a scheme has been introduced in the beginning of the twentieth century in the framework of the study of "singularities of elastic media". It has preceded the crystallographic description given in the previous sections.

A "Volterra dislocation" is a singular line, involving a discontinuous displacement, within an elastic medium. It is formed by a procedure similar to that used in the crystallographic description. To clarify this analogy let us start with the cases of straight dislocations of the edge and screw types.

5.4.1 *Edge and screw dislocations*

Edge-dislocation formed by slipping

In a *continuous elastic medium*, consider (Fig. 5.12 (left)) the straight line D and a horizontal half-plane limited by D. Make a cut of the medium along this half-plane. One thus defines two "lips". The lower one S borders the lower half of the crystal, and the upper one is adjacent to the upper half of the crystal. Displace S' by sliding it uniformly (except in the vicinity of D which is a line of discontinuity of the displacement) by a translation $\vec{b}$ parallel to the plane of the cut, and perpendicular to D, the modulus of $\vec{b}$ being arbitrary.

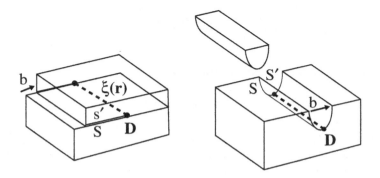

Figure 5.12   Formation of an edge-dislocation in a continuous medium by slipping or by insertion of matter.

After this displacement, the continuity of the medium is restored by "gluing" together the surfaces facing eachother.[10] The external forces which have imposed the relative displacement of the surfaces are then suppressed and the medium is left to regain its elastic equilibrium. This process induces a displacement field $\overrightarrow{\xi}(\overrightarrow{r})$ in the entire elastic medium which is *discontinuous across* D. This line is an edge dislocation of the elastic medium. Note that, in the crystallographic definition of a dislocation, the preservation of the crystal structure outside the vicinity of D requires that b be equal to a lattice translation. Here, the assumed continuity of the medium allows an arbitrary value for b (assumed, however, to be large with respect to the primitive translations of the crystal, owing to the fact that the scale of the description is much larger than the atomic scale).

Edge dislocation generated by adding or removing matter

If one chooses a cut along a half plane which does not contain $\overrightarrow{b}$, an edge dislocation can be formed according to different processes involving the addition or removal of matter. These processes will be the counterparts of the crystallographic formation of an edge-dislocation by insertion or removal of one or several lattice half-planes (cf. section 5.1). For instance, if $\overrightarrow{b}$ is perpendicular to the cut (Fig. 5.12 (right)) the translation of S' with respect to S, in the direction indicated, will induce a "trench" in the solid. In order to form a dislocation, one can "fill" this trench by inserting a layer of matter of thickness b, and then proceed with the gluing of this layer to the two surfaces S and S'. After suppression of the external forces which have created the trench, the solid will reach a new elastic equilibrium, involving a deformation field.

If, for the same set of surfaces S and S' the translation $\overrightarrow{b}$ has the opposite sense (then tending to push the matter on either sides to overlap) one has, in order to achieve the displacement, to *remove* a layer of matter of thickness b.

Other orientations of the half planes S and S' (neither parallel, nor perpendicular to $\overrightarrow{b}$) will determine intermediate situations in which an edge dislocation will be formed both by a relative slipping of planes, and by addition or removal of matter. As for crystallographically-defined dislocations, the nature of the edge dislocation thus formed will not depend of the

---

[10]The preceding "gluing operation" is equivalent to the restoration of the chemical bonds between neighbouring atoms which is performed in the crystallographic procedure.

procedure but only of the relative directions of D (the defect line) and of the displacement $\vec{b}$ (which determines the discontinuity across D). Indeed, only those two elements remain after restoration of the continuity of the medium, and suppression of the external forces.[11] This independence with respect to the procedure of formation will be confirmed, in chapter 6, section 6.2), by the result of the calculation of the deformation field associated to the dislocation.

Screw dislocation

In a similar manner, consider an infinite line D bordering a half plane (Fig. 5.13 (left)). A cut along this half-plane creates two surfaces S and S'.Impose a uniform slip of one of the surfaces of amplitude $\vec{b}$ parallel to D, excluding the vicinity of D. The subsequent gluing together of the surfaces facing eachother, and the suppression of the external shear forces which have produced the relative slip of the surfaces, will restore the continuity of the medium and form a *screw* dislocation D. A displacement field $\vec{\xi}(\vec{r})$, which is discontinuous across D, is thus generated in the entire medium.

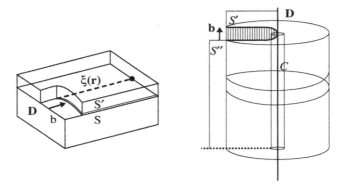

Figure 5.13   Two representations of the formation of a screw dislocation in a continuous medium.

As shown on Fig. 5.13 (right), a set of circles perpendicular to D, and regularly spaced at a distance , along D, is transformed, after formation of

---

[11]Since the medium is assumed to be infinite, the step of amplitude $\vec{b}$ represented on the figure, for the sake of clarity, has no real existence.

the dislocation, into a helix of pitch b. The displacement $\xi(\theta)$ of a point is such as:

$$I = \xi_z(\theta + 2\pi) - \xi_z(\theta) = b \qquad (5.3)$$

A loop (L) is the equivalent of a Burgers circuit (cf. section 5.2.2) and $\vec{b}$ is the Burgers vector of the dislocation, equal to the magnitude of the slip imposed to form the dislocation. Note, on the basis of Fig. 5.13 (right), that the deformation induced does not depend of the choice made for the half-plane bordered by D.

5.4.2 *General case*

To generalize the preceding definitions, consider an elastic medium and C a closed loop (not necessarily contained in a plane). Let S be a portion of surface limited by C (Fig. 5.14). A cut of the medium along S will define two surfaces S' and S, facing eachother. Displace S' by a uniform translation $\vec{b}$, excluding from the displacement a neighbourhood of C defined by a cylinder surrounding C and of radius approximately equal to b. Assuming that, outside the space between S and S', the continuity of the medium is preserved, such a "forced" displacement of S', will induce a displacement field in the entire elastic medium, C being a line of discontinuity.

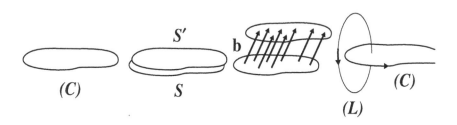

Figure 5.14   Volterra procedure of formation of a dislocation along line (C). From left to right: Creation of a dislocation loop in a continuous medium. Cut decomposing the loop into two "lips". Displacement (translation) of the upper lip, which creates a void between the lips. This void is filled with matter identical to the surrounding medium. Burgers circuit surrounding the dislocation loop.

Depending of the direction of $\vec{b}$ with respect to S, and of the shape of S, the translation of S' can be a slipping of S' on S, or a displacement which creates a void in the elastic medium. It can also be such as to push the matter above S' inside the matter below S. In the two latter cases, one has either to fill the void by a same volume of the elastic medium, or withdraw the excess matter preventing the required displacement.

In all cases, the continuity of the medium is restored, after the former operations, by gluing the surfaces facing eachother. Subsequently, the external forces which have permitted the displacement are suppressed and a new elastic equilibrium is established in the medium. In this system, C is a dislocation. Indeed, adjacent elements of matter on both sides of C have undergone either a displacement $\overrightarrow{b}$, or a zero displacement. As noted, the initial elastic medium is deformed by a displacement field $\overrightarrow{\xi}(\overrightarrow{r})$. If a sense is chosen on C, a right handed orientation is defined for the loops (L) surrounding C. One then has:

$$I = \oint_L \frac{\partial \overrightarrow{\xi}}{\partial l}.dl = \overrightarrow{b} \qquad (5.4)$$

Any loop (L) is the equivalent of a Burgers circuit. (cf. section 5.2.2) and $\overrightarrow{b}$ is the *Burgers vector of the dislocation*, equal to the displacement imposed during the formation of the dislocation.

As in the case of their crystallographic counterpart, the nature of the dislocations in continuous media (as well as their mechanical effects) do not depend on the choice of the surface S within the medium, but only of the shape of the curve C and of the displacement $\overrightarrow{b}$, since only the two latter elements remain after restoration of the continuity of the medium.

5.5   Observation of dislocations

Direct experimental observations of dislocations, at the microscopic scale, and the determination of their characteristics could be first achieved around 1960, using *electronic microscopy*. These observations constituted an important evidence of the role of dislocations in the mechanism of the plastic deformation. This method is still today the main mean of analysis, at the atomic or mesoscopic[12] scale, of the mechanical properties of solid materials. Therefore, it is useful to understand the principle of the visualization of dislocations by means of the electron microscope.

---

[12]The mesoscopic ("intermediate") scale concerns properties which vary on distances comprised between a few nanometer and a few microns.

5.5.1 *Reflexion of electrons by a crystal*

An electron microscope partly operates in a similar way as an optical microscope: exposure of the object (a small crystal) to a parallel beam of monokinetic electrons (i.e. all having the same velocity), scattering of the electrons by the different regions of the object, formation of successive images with help of magnetic "lenses" (Fig. 5.15).

Quantum mechanics ensures that the incident electron beam can be considered as a monochromatic plane wave whose wavelength is the de Broglie wavelength $\lambda = (h/p)$ associated to the momentum $p = m\mathbf{v}$ of the electrons. In an electron microscope, electrons are accelerated by a voltage of a few hundred kilovolts, and their associated wavelength is of the order of 10^{-2}Å. The smallness of such a wavelength, as compared to the interatomic distances in a crystal (of the order of one Angström) plays a central role, as will be seen, in the practical procedure of observation of dislocations.

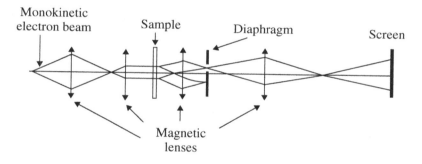

Figure 5.15   Schematic configuration of an electron microscope.

Bragg rule

As shown in chapter 2 (cf. section 2.2) the Bravais lattice of a crystal can be considered as a stacking of parallel, and equally spaced, lattice planes. The principles of the scattering of an electron beam by a crystal, as well as the observation of dislocations, can be deduced from this description.

Indeed, assume that the plane wave associated to the electron beam is incident, under the angle θ on a set of N, parallel, and adjacent, lattice planes, equally spaced by a distance d (Fig. 5.16). Due to the interaction of the wave with the matter contained in these planes (nuclei and electrons), the wave will be reflected by each plane. In a simplified approach, assume that the fraction R of the wave amplitude which is reflected is sufficiently small (e.g. $R << 1$) to ensure that all the successive planes are submitted, approximately, to the same wave amplitude. This smallness also allows to neglect the secondary reflection of a reflected wave on another adjacent planes (no "multiple" reflections). On the other hand, two waves reflected by adjacent planes have the same amplitude, and a phase shift equal to:

$$\phi = \frac{2\pi}{\lambda}(2d\sin\theta) \tag{5.5}$$

The amplitude of the *scattered beam* is the sum of all the reflected waves on the set of N planes:

$$A \propto R(1 + e^{i\phi} + ... + e^{i(N-1)\phi}) \tag{5.6}$$

Generally, the number N is large: for instance, for a crystal thickness of the order of one micron, one has $N \approx 10^3$. If the phase shift ϕ is, even slightly, different from 2π, *the amplitude A of the reflected wave is equal to zero,* as shown by Fig. 5.16 which represents the sum of the complex amplitudes of N successive waves, each one being shifted by ϕ with respect to the preceding one.

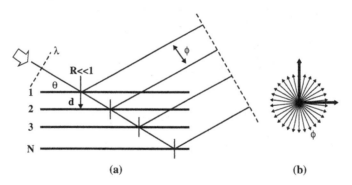

(a)   (b)

Figure 5.16   (Left) Reflection of a plane wave on a set of parallel lattice planes. (Right) Schematic representation of the sum, in the complex plane, of waves possessing a relative phase shift of arbitrary value between successive waves.

By contrast, if $\phi = 2h\pi$, the amplitudes of the successive waves add to eachother. Although each wave has, as assumed, a very small amplitude, their number N is large enough to result in a scattered beam having a significant fraction of the incident amplitude.

In conclusion, an electron beam incident on a set of lattice planes under an arbitrary angle θ, will not generate any scattered beam, unless the angle of incidence obeys the so-called Bragg rule:

$$\frac{\lambda\phi}{2\pi} = 2d\sin\theta_d = h\lambda \tag{5.7}$$

Since the different sets of lattice planes have definite orientations with respect to the crystal (i.e. with respect to the surfaces limiting the crystal sample), this rule has the consequence that if the crystal has an arbitrary orientation with respect to the electron beam, no scattering will generally be observed. A scattering will be generated only if the the beam is at a Bragg incidence on one set of lattice planes.

5.5.2 How can a dislocation be observed?

Deformation of lattice planes by a dislocation

Figure 5.17 shows a cystal containing a straight edge-dislocation D . This dislocation borders an additional half-atomic plane. Consider the set of lattice planes parallel to this half plane (drawn horizontal on the figure) and an electron beam incident on it at the Bragg incidence θ_D, such as $2d\cos\theta_D = h\lambda$. The set of lattice planes being deformed in the neighbour-hood of the dislocation, the distance between adjacent planes differs from d. Hence, the Bragg rule will not be obeyed in this region (surrounded by a circle), and therefore, the neighbourhood of the dislocation will not contribute to the scattered beam.

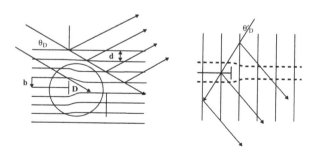

Figure 5.17   (Left) Reflection of a beam falling, at Bragg's incidence, on lattice planes perpendicular to $\vec{b}$. (Right) Reflection at Bragg's incidence, on lattice planes parallel to $\vec{b}$.

The considered set of planes is the set most deformed by the disloca-
tion. Note that it is *perpendicular to the Burgers vector* $\overrightarrow{b}$ of the edge-
dislocation. The same remark holds for a screw-dislocation for which the
most deformed planes form a helicoidal ramp perpendicular to $\overrightarrow{b}$.

By contrast, the B sets of planes parallel to the dislocation, and whose
normal vector $\overrightarrow{n}$ is perpendicular to the Burgers vector (drawn vertical on
the figure) are not deformed near D. Hence, an electron beam falling at
Bragg's incidence on this set of plane will be scattered in the entire volume
of the crystal including the neighbourhood of the dislocation.

The sets of lattice planes having different orientations have a variable
degree of deformation, more pronounced if $\overrightarrow{n}$ has a significant projection
on $\overrightarrow{b}$.

Imaging a dislocation

In optics, an image of a planar object can be formed with help of a lense by
illuminating the object with a parallel beam of light and by placing a screen
in a position "conjugated" to that of the object with respect to the lense.
On the screen, the image is constituted by light scattered by the different
points of the object.

Likewise, in electron microscopy, the object, a thin slice of a crystal, and
a "fluorescent" screen are at conjugated positions with respect to a magnetic
lenses $L1$, thus permitting the formation of an image of the crystal slice.

However, an important specificity of electron microscopy, resides in the
fact that *the incident beam must be oriented with respect to the crystal, at
a Bragg incidence for a set of lattice planes.* Otherwise, the crystal will not
reemit the beams used to form the image. *This remark holds for each point
of the crystal.*

In order to visualize a dislocation one has to form an image of the
dislocated crystal with help of an electron beam oriented at the Bragg incidence θ_D *relative to a set of lattice planes deformed by the dislocation.*
Consider, for instance, the set of lattice planes denoted A. In the "good
crystal" regions, the electron beam is assumed to be at the Bragg incidence. These regions scatter efficiently the incident beam. All the beams,
scattered by the various regions of the crystal, are made to converge on
the screen by the magnetic lenses and form corresponding images of these
crystal regions, which will appear illuminated on the fluorescent screen. By
contrast, in the region of the core of the dislocation, the Bragg rule being
in default, no reflected beam is generated. No beam passes in the conju-

gated region of the screen: the image of the dislocation core appears as a dark line.

Hence, the image of a crystal, obtained with an electron beam at the incidence θ_D will appear as a dark line within a clear surrounding crystal.

If, instead, an image of the crystal is formed by using an electron beam oriented at a Bragg incidence on a set of lattice planes which are not (or little) deformed by the dislocation, i.e. such as $\vec{n}.\vec{b} \approx 0$ (for instance the B planes in Fig. 5.17) the dislocation will not be visible because all the regions of the crystal (including the dislocation-core) will be simultaneously at the Bragg incidence.

Determination of the Burgers vector

The preceding considerations also provide a method for determining experimentally the direction of the Burgers vector $\vec{b}$. Indeed, in this view, one must form the image of the crystal, using successively electron beams reflected on different sets of lattice planes. Let $\vec{n}_i$ be the normals to these different sets of planes. The dislocation will be visible if $\vec{n}_i.\vec{b} \neq 0$ and it will not be visible if $\vec{n}_i.\vec{b} = 0$. Hence a series of observations will generally permit a non-ambiguous specification of the *direction* of $\vec{b}$. As will be shown in the next chapter, the modulus of $\vec{b}$ can be taken as the smallest lattice translation in this direction. The prior knowledge of the crystallographic characteristics of the considered solid will allow to deduce from the observations the value of $\vec{b}$.

It is worth noting that, since the local direction of the dislocation appears on the image, one is then able to determine the relative orientations of $\vec{b}$ and of the dislocation-line, and deduce the edge-, screw-, or mixed-, nature of the dislocation.

5.5.3 *Lattice planes and reciprocal lattice*

As mentioned in chapter 2, a crystal contains an infinite (though discrete) number of differently oriented sets of lattice planes. Each direction is defined by a choice of three nodes of the Bravais lattice. Generally different sets of equally spaced planes are associated to different distances d between adjacent planes. If d is very small, the corresponding set of planes cannot satisfy the Bragg rule. Indeed the Eq. (5.7) expressing this rule implies that the wavelength λ of the electron beam be smaller than $2d$. By taking the three nodes of the Bravais lattice defining a plane, sufficiently distant from eachother, d can be made arbitrarily small.[13] Hence, *only a finite number of lattice plane orientations will be liable to satisfy the Bragg rule*, or, in other terms, a finite number of relative orientations of the incident electron beam with respect to the crystal will determine the occurence of reflected electron beams. However, their number is sufficiently large to require an analysis of their specification and of their selection.

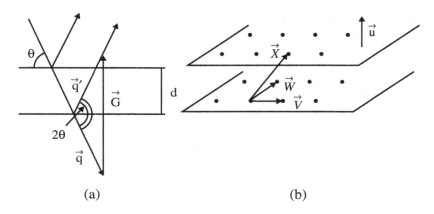

(a)   (b)

Figure 5.18   (a) Vectorial form of the Bragg rule. (b) Primitive translations adapted to a direction of lattice planes.

Reciprocal lattice

A set of parallel and equally spaced lattice planes can be defined by its normal $\vec{n}$ and the value of the distance d between adjacent planes. Let $\vec{g} = (2\pi/d)\vec{n}$. The knowledge of $\vec{g}$ defines entirely the set of lattice planes.

The directions of the different sets of lattice planes of a crystal are related to the characteristics of its primitive translations $\vec{a}_i$. Indeed these planes join nodes of the Bravais lattice generated by these translations. The precise relationship can be deduced from a specific choice of primitive translations. For a given direction of lattice planes, two of the translations

---

[13] An additional reason for the irrelevance of certain plane directions, even if their distance is sufficient to satisfy the Bragg rule, is that these planes, being defined by distant nodes, will possess a small density of nodes, hence a small density of matter (i.e. atoms). Their "reflection" coefficient R, resulting from the interaction of electrons with the corresponding atoms, will be small, thus determining a weak contribution to the global reflection.

can be taken to join neighbouring nodes within one of the lattice planes, while the third one joins any two nodes of adjacent planes of the set (cf. Fig. 5.18 (right)). On the basis of this choice, it can be shown, that any $\vec{g}$ vector defining a set of lattice planes can be expressed as:

$$\vec{g} = m_1 \vec{a}_1^* + m_2 \vec{a}_2^* + m_3 \vec{a}_3^* \tag{5.8}$$

in which the m_i are integer coefficients and in which the $\vec{a}_i^*$ vectors are related to the primitive translations $(\vec{a}_1, \vec{a}_2, \vec{a}_3)$ by:

$$\vec{a}_i^* = 2\pi \frac{\vec{a}_j \wedge \vec{a}_k}{(\vec{a}_1, \vec{a}_2, \vec{a}_3)} \tag{5.9}$$

Equation (5.8) shows that the set of end points of the $\vec{g}$-vectors generate a three dimensional periodic set of points forming a "lattice" whose basis is constituted by the $\vec{a}_i^*$. This lattice is called the *reciprocal lattice* of the crystal (the Bravais lattice generated by the $\vec{a}_i$ is often called the "direct lattice").

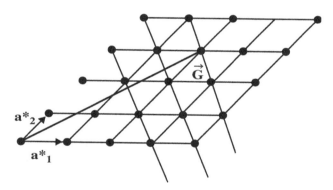

Figure 5.19   Reciprocal-lattice plane.

Similarly to the direct lattice, the reciprocal one can be considered as a stacking of parallel, equally spaced "reciprocal lattice planes". One such plane P, contains a regular lattice of nodes (Fig. 5.19), end points of vectors denoted $\vec{G}$. A given node will define a direction of "direct" lattice planes whose normal vector $\vec{n}$ is parallel to $\vec{G}$, while the distance d between adjacent planes is given by $d = 2\pi/|\vec{G}|$. Since the directions of the $\vec{a}_i^*$ are determined by the directions of the $\vec{a}_i$, a rotation of the crystal, which acts on the $\vec{a}_i$ will also rotate the $\vec{a}_i^*$ as well as the reciprocal lattice planes.

Vector formulation of the Bragg rule

Figure 5.18 (a) shows that if $\overrightarrow{q}$ and $\overrightarrow{q'}$ are the wavevectors, repectively, of the incident, and of the reflected waves (with $|\overrightarrow{q'}| = |\overrightarrow{q}| = (2\pi/\lambda)$), then $(\overrightarrow{q'} - \overrightarrow{q})$ is parallel to the normal $\overrightarrow{n}$ to the lattice plane (hence parallel to $\overrightarrow{g}$). Besides the modulus $|\overrightarrow{q'} - \overrightarrow{q}|$ is equal to $(4\pi \sin\theta/\lambda)$. Consequently, the Bragg rule, which can be written as $(4\pi \sin\theta/\lambda) = h.(2\pi/d)$, is equivalent to:

$$(\overrightarrow{q'} - \overrightarrow{q}) = h\overrightarrow{g} = \overrightarrow{G} \tag{5.10}$$

in which $\overrightarrow{g}$ défines a set of lattice planes. Obviously, (cf. Eq. (5.8)), $\overrightarrow{G}$, a multiple of $\overrightarrow{g}$, is also a reciprocal lattice vector.

5.5.4   *Diffraction of electron beams*

As seen in the preceding section, determining the characteristics of a dislocation requires obtaining several images of the dislocation using reflections on different sets of lattice planes (i.e. implying several Bragg incidences). In principle this would lead to adjust, for each image, the orientation of the crystal with respect to the incident electron beam (which has a fixed direction parallel to the axis of the microscope). However, this is not necessary, in general, because *it is possible to realize simultaneously the reflection of the incident electron beam on several, differently oriented sets of lattice planes.*

This favourable situation is due to the smallness of the electron wavelength λ with respect to the interplanar distances d. These distances are of the order of one Angström, while $\lambda \approx 10^{-2}$ Angströms, as mentioned earlier. As a consequence, the modulus $|\overrightarrow{q}|$ of the incident and reflected wavevectors (which are equal) are such as $|\overrightarrow{q}| = (2\pi/\lambda) >> |\overrightarrow{G}|$. The angle 2θ between $\overrightarrow{q}$ et $\overrightarrow{q'}$ is very small and thus (Fig. 5.20) $\overrightarrow{G}$, is approximately perpendicular to $\overrightarrow{q}$ and $\overrightarrow{q'}$.

In practice the orientation of the crystal will be set in such a way as to have a reciprocal lattice plane P perpendicular to $\overrightarrow{q}$. Chosing the origin[14] of this plane at the end-point of $\overrightarrow{q}$, the Bragg rule $(\overrightarrow{q'} - \overrightarrow{q}) = \overrightarrow{G}$ will also be simultaneously obeyed[15] for all the lattice plane directions associated

---

[14] As for the direct Bravais lattice, the reciprocal lattice has an arbitrary origin.

[15] It is only aproximately obeyed but the studied crystals being very thin, the number N of "direct" lattice planes contained in its thickness is not very large and one can show that this implies a certain angular tolerance $\Delta\theta$, in the angular condition set by the Bragg rule.

Figure 5.20   Bragg rule geometry (a) in the general case, and (b) when the wavelength of the incident beam is small as compared to the interatomic spacing.

with the vectors $\vec{G}_1$, $\vec{G}_2$, $\vec{G}_3$, ... Consequently, several diffracted beams, with wavevectors $\vec{q}_1$, $\vec{q}_2$, $\vec{q}_3$, will be simultaneously emitted (Fig. 5.21).

Selection of a given diffraction direction

As emphasized, the possibility of a simultaneous diffraction of the incident electron beam on several sets of lattice planes has the advantage of avoiding a reorientation of the crystal between successive images. It has the draw-back, if all the diffracted beams are simultaneously used to form an image of the crystal, to superimpose images involving lattice planes more or less deformed by the dislocation. The contrast between the dislocation and the surrounding crystal will be weakened. It is possible to obtain an optimum contrast, without modifying the orientation of the crystal by selecting a single diffracted beam.

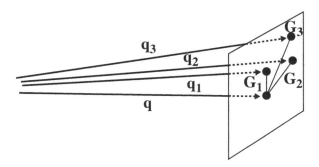

Figure 5.21   Crystal oriented such as to position a reciprocal plane normal to the wavevector of the incident beam. The Bragg rule is simultaneously obeyed for several $\vec{G}_i$ vectors, thus determining several scattered beams.

In this view, an additional magnetic lens L can be placed after the crystal (Fig. 5.15) perpendicular to $\vec{q}$. The different diffracted beams, having distinct directions will converge at different points of the focal plane of L. By comparing Figs. 5.15 and 5.21, it appears, in fact, that this focal plane displays a distribution of intensities which maps the reciprocal plane P containing $\vec{G}, \vec{G}_1, \vec{G}_2, \vec{G}_3, \ldots$ In order to select a single direction of diffracted beam, it is sufficient to place an opaque screen in this focal plane with a small aperture (a diaphragm) surrounding the location of one only of the preceding $\vec{G}$ vectors, e.g. $\vec{G}_1$. Only the corresponding beam (with wavevector $\vec{q}_1$ will propagate beyond the screen. Finally, with help of a third lens L' an image of the crystal will be formed with the sole contribution of this beam.

Exploring the various images formed by different diffracted beams, will then consist in displacing the diaphragm in the focal plane of L, isolating each time a different "spot" $\vec{G}_i$, and observing, on the final screen, the contrast in the corresponding image of the crystal (cf. exercise 11). Figure 5.22 provides an example of a dislocation loop revealed by electron microscopy.

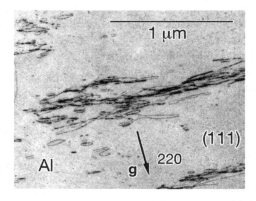

Figure 5.22   Dislocation loops in an aluminum crystal submitted to a 2% shear strain at a temperature of 77K. The center of the image shows an accumulation of loops close to each other (Courtesy of Patrick Veyssière ONERA).

Chapter 6

Strain field of dislocations

Main concepts: Strains and stresses induced by straight dislocations. Infinite energy per unit length associated to a straight dislocation. Effective force, induced by a stress field, acting on a dislocation. Line tension of a dislocation line. Interaction between parallel dislocations. Interaction between a dislocation and a vacancy.

6.1 Introduction

It has been shown qualitatively in the preceding chapter (section 5.2.2) that a dislocation creates in the surrounding crystal a strain field which decreases slowly (b/r law) with the distance r to the dislocation. This strain can be considered as the result of an "internal stress" produced by the dislocation at each point of the crystal. These "internal" strain and stress are the keys to understand two central features of the plastic behaviour of a crystalline solid.

On the one hand, acting on a crystal with an *external* stress, will modify the strain field (by adding an external component to the internal one). The dislocation will "feel" this change of the deformation field in its surrounding as a "force" putting the dislocation into motion. As mentioned in the overview given in the introductory chapter, it is this very motion which constitutes the elementary step of the plastic deformation.

On the other hand, the internal stress determined by a dislocation will modify the strain field associated to other defects present in the solid (e.g. other dislocations, vacancies or other point defects, surfaces or grain boundaries...). It will therefore exert a force on these defects and, conversely the defects will exert forces on it. These interactions will have the effect, either to make more easy the motion of the dislocation, or to oppose it, thus making easier or more difficult the plastic deformation.

Due to the slowness of the decrease of the deformation as a function of the distance to the dislocation line, the characteristic scale of its spatial variations will be large as compared to b, hence to the interatomic distances. In studying more precisely these variations, it is therefore justified to consider that the crystal is a *continuous elastic medium* (with the exception of the dislocation core). In such an approximation, it is consistent to define a dislocation by the corresponding Volterra procedure (cf. section 5.4).

Besides, the approximate amplitude of the strain ($\epsilon \approx (b/2\pi r)$) is small even near the core of the dislocation ($r = 5b \rightarrow \epsilon < 5\%$). For this range of strain values, the equations of *linear elasticity (section 3.2)* will apply.[1] In order to simplify the calculations we will assume, in addition, that the considered solid is an *isotropic* elastic medium.

In this framework, four subjects will successively be dealt with.

In a first step, the strain and stress fields induced by the presence of a dislocation will be analyzed, as well as the mechanical energy associated to these fields. This analysis will be restricted to the cases of straight dislocations for which simple expressions are obtained. The case of loop dislocations will only be discussed qualitatively.

It will then be shown that an external stress field acting on a solid containing a dislocation exerts an effective force on the dislocation whose expression will be derived.

Considering the amount of mechanical energy associated to a dislocation will then lead to define the *line tension per unit length* of a dislocation, which is a force tending to favour the straight-line shape of a dislocation.

Finally, the interaction between two dislocations as well as the interaction between a dislocation and a point defect will be examined.

6.2 Strain and stress fields

In the absence of external stresses, an isotropic continuous solid containing a dislocation is in elastic equilibrium. The strain and stress fields can be deduced from a displacement field $\vec{\xi}(\vec{r})$ satisfying the equilibrium equation (3.30):

$$(1 - 2\nu)\Delta\vec{\xi} + \vec{\nabla}(div\,\vec{\xi}) = 0 \qquad (6.1)$$

[1]A strain value of 5% is larger than the *measured value* of the limit of the elastic behaviour in most solids ($\epsilon < 10^{-3}$) However, one must have in mind, as will be shown in chapter 7, that this *measured value* is related to the motion of dislocations. In the problem considered here, the relevant value corresponds to the solid *external* to the dislocation (hence free from dislocations). The value of the limit is then generally larger than 5%.

in which ν is the Poisson coefficient (3.31). In addition, $\vec{\xi}(\vec{r})$ has to comply with the following conditions:

a) $\vec{\xi}(\vec{r})$ is only defined outside the dislocation-core, i.e. a singular region of radius r_0 surrounding the dislocation line.

b) Integrated along a Burgers circuit around the dislocation, $\vec{\xi}(\vec{r})$ has a defect of closure (Eq. (5.4)):

$$\oint \frac{\partial \vec{\xi}}{\partial L} dL = \vec{b} \tag{6.2}$$

c) The strain and stress fields deriving from $\vec{\xi}(\vec{r})$ must have zero values at infinity, for an infinitely extended solid. Alternately, the stress field must have zero value, on the outer surface limiting a solid of finite size.[2]

Also note that, for straight dislocations in an infinite solid, the various fields considered are uniform in the direction parallel to the dislocation-line. Hence, a cylindrical frame of coordinates (cf. Table 3.1) is adapted to express the required solutions of Eq. (5.4).

6.2.1 Screw dislocation

Displacement field

Figure 6.1 represents a straight, right-handed, screw dislocation in an infinite elastic medium. The z-axis is chosen along the $\vec{d}$ vector which orients the dislocation line. It has the same direction and sign as the Burgers vector $\vec{b}$ (for a left-handed screw dislocation $\vec{b}$ would be directed opposite to $\vec{z}$).

As emphasized in chapter 5, a screw dislocation does not involve additional or missing half lattice planes, and therefore it neither induces local compressions or dilatations in the solid. This will be consistent with solutions of the equilibrium equation for which the local volume of matter is preserved, i.e. satisfying the condition $div\,\vec{\xi} = \sum \epsilon_{ii} = 0$ (cf. section 3.2.1). The equilibrium equation then reduce to:

$$\Delta \vec{\xi} = \left(\frac{\partial^2}{\partial r^2} + \frac{1}{r}\frac{\partial}{\partial r} + \frac{1}{r^2}\frac{\partial^2}{\partial \theta^2} + \frac{\partial^2}{\partial z^2} \right)\vec{\xi} = 0 \tag{6.3}$$

---

[2]As shown in exercise 16, different types of boundary conditions can be relevant on external surfaces depending on their "rigid" or "deformable" character.

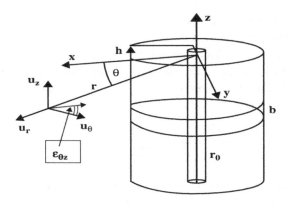

Figure 6.1   Cartesian and cylindrical frames adapted to the study of a screw-dislocation. The only non zero strain-component is the shear component $\epsilon_{z\theta}$ corresponding to the generation of a helix along z.

In addition, the known characteristics of a screw dislocation impose: i) That $\vec{\xi}$ be independent of the variable z. ii) That $\xi_z(r, \theta+2\pi) = \xi_z(r, \theta)+b$. The following solution:

$$\xi_x = \xi_y = 0 \qquad \xi_z = \frac{b}{2\pi}\theta = \frac{b}{2\pi}\arctan(\frac{y}{x}) \tag{6.4}$$

has zero divergence and satisfies conditions i) and ii).   Note that the polar and cartesian formulations in 6.4 are not strictly equivalent since $\arctan(y/x)$ takes values in the interval $(\pi/2, +\pi/2)$. This mathematical difference has no consequence on the value of the strain field which is only a function of the gradient of ξ_z.

Strain and stress fields

The derivation from expression (6.4) of the cartesian components of the strain tensor $\bar{\bar{\epsilon}}$ through equation (Eq. (3.1)) shows that the only non-zero components are:

$$\epsilon_{13} = \epsilon_{31} = \frac{-by}{4\pi r^2} = \frac{-b\sin\theta}{4\pi r} \qquad \epsilon_{23} = \epsilon_{32} = \frac{bx}{4\pi r^2} = \frac{b\cos\theta}{4\pi r} \tag{6.5}$$

Table 3.1 in chapter 3 then shows that a single component $\epsilon_{\theta z} = \epsilon_{z\theta}$ is non-zero in cylindrical coordinates:

$$\epsilon_{\theta z} = \frac{b}{4\pi r} \tag{6.6}$$

Physically, this expression means that the (initially) right-angle between the vectors $\vec{z}$ and $\vec{u}_\theta$ (Fig. 6.1) becomes $(\pi/2 - \delta\phi)$ with $\delta\phi = 2\epsilon_{\theta z} = (b/2\pi r)$. This is consistent with the fact that a right-handed screw dislocation will transform circles perpendicular to the dislocation line, into a helix of pitch b (cf. section 5.4.1).

From relation (3.26) pertaining to an isotropic elastic medium, we can then infer that, since the considered strain component is a pure shear, the stress tensor has components which are proportional to the components of the strain tensor, the proportionality coefficient being 2μ:

$$\sigma_{13} = \frac{-\mu b \sin\theta}{2\pi r} \qquad \sigma_{23} = \frac{\mu b \cos\theta}{2\pi r} \qquad \sigma_{\theta z} = \frac{\mu b}{2\pi r} \qquad (6.7)$$

In agreement with the qualitative analysis in section 5.2.2, it appears that the strain and stress fields decrease as $(1/r)$ and are proportional to the modulus b of the Burgers vector.

Elastic energy density

Application of the general expression (eq.3.5) of the elastic energy density, using the components (6.6) and (6.7) yields :

$$u = \frac{1}{2}[\sigma_{\theta z}\epsilon_{\theta z} + \sigma_{z\theta}\epsilon_{z\theta}] = \frac{\mu b^2}{8\pi^2 r^2} \qquad (6.8)$$

Consider the volume comprised between the core (cylinder of radius r_0 surrounding the dislocation) and a cylinder of large radius R. A *unit height* of this volume contains a total energy:

$$\frac{dU(R)}{dL} = \frac{\mu b^2}{8\pi^2} 2\pi \int_{r_0}^{R} \frac{r dr}{r^2} = \frac{\mu b^2}{4\pi} Ln(\frac{R}{r_0}) \qquad (6.9)$$

The integral in the second member becomes infinite for $R \to \infty$. The above formula therefore shows that the *elastic energy per unit length* induced by the dislocation in the surrounding infinite solid (in the radial direction) *is infinite*. This result contrasts with the one found for a vacancy cf. Eq. (4.12), for which the energy created is finite.

The preceding expression (6.9) can be alternately deduced from the amount of work needed to form a screw dislocation by the "slipping" procedure (section 5.3.1). Indeed, at a distance r of the dislocation, the value of the stress which has to be used in order to induce a relative slipping αb ($0 \le \alpha \le 1$) of the planes facing eachother along the cut, is $\sigma_{\theta z} = (\mu\alpha b/2\pi r$. On an infinitesimal surface of unit-height and of width dr, the force corresponding to this stress is $df = \sigma_{\theta z}.dr$. Hence, the total work needed to form a unit height of the dislocation is:

$$\int \int df.bd\alpha = \frac{\mu b^2}{2\pi} \int_{r_0}^{R} \frac{dr}{r} \int_0^1 \alpha d\alpha = \frac{\mu b^2}{4\pi} Ln(\frac{R}{r_0}) \qquad (6.10)$$

Instability of dislocations

Let us assume that the preceding expressions of the stress field remain valid for a finite crystal sample,[3] having the shape of a cylinder of radius $R \approx 1\text{cm}=10^8\text{Å}$. Take for instance $b = 3\text{Å}$, $r_0 \approx 2b$ and $\mu \approx 5.10^{10}\text{J/m}^3$. For a length of the dislocation equal to one atomic diameter, i.e. $h \approx b \approx 3\text{Å}$, The elastic energy will be ≈ 3 eV.[4] It can also be shown that the entropy associated to the presence of dislocations is such as $TS << U$, in contrast with the case of vacancies (cf. chapter 4). As consequence, *a dislocation is never thermodynamically stable*. In a crystal, the increase of energy brought by the dislocation (very high even in a finite crystal) is not compensated by an entropy increase. The occurence of dislocations in a real crystal corresponds to a metastable state, its origin being related to the mechanical and thermal "history" of the sample.

6.2.2   Straight edge dislocation

Displacement field

Let us choose the cartesian frame of reference of an edge dislocationas defined in section 5.2.2, and recalled in Fig. 6.2. The displacement field of this dislocation must comply with conditions similar to those imposed to a screw dislocation (cf. section 6.2.1), $\overrightarrow{b}$ being now directed along x. In contrast with a screw dislocation, an edge-one is associated to a region of compression (in the direction of the half-plane $y > 0$) and a region of expansion (in the direction of the missing half-plane $y < 0$). As a consequence, the displacement field has $div(\overrightarrow{\xi}) \neq 0$. The direction of $\overrightarrow{b}$ suggests to look for a solution of the equilibrium Eq. (6.1) independent of the coordinate z:

$$\xi_z = 0 \qquad \frac{\partial \xi_x}{\partial z} = \frac{\partial \xi_y}{\partial z} = 0 \qquad (6.11)$$

---

[3]This assertion is not necessarily true: for a finite crystal the stress field could be modified by the boundary conditions on the surface.

[4]One must add to this value the *core energy*. Available numerical simulations show that, in metals, this energy is of the order of a fraction of electron-volt (cf.exercise 18).

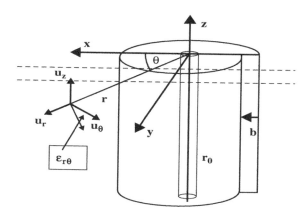

Figure 6.2 Cartesian frame (x, y, z) and cylindrical frame (u_r, u_θ, u_z) for an edge-dislocation. The deformation induced consists of a compression or an expansion parallel to u_r and u_θ, as well as shear $\epsilon_{r\theta}$.

The derivation of this solution, which derives from general methods of study of continuous elastic media, will not be reproduced here. It has the form:

$$\xi_x = \frac{b}{2\pi} \left[\arctan(\frac{y}{x}) + \frac{xy}{2(1-\nu)\mathbf{r}^2} \right] = \frac{b}{2\pi} \left[\theta + \frac{\sin 2\theta}{4(1-\nu)} \right] \qquad (6.12)$$

$$\xi_y = \frac{-b}{8\pi(1-\nu)} \left[2(1-2\nu)Ln(\mathbf{r}) + \frac{(x^2-y^2)}{\mathbf{r}^2} \right] \qquad (6.13)$$

$$= \frac{-b}{8\pi(1-\nu)} \left[2(1-2\nu)Ln(\mathbf{r}) + \cos 2\theta \right] \qquad (6.14)$$

Strain and stress fields

Owing to the planar character of the displacement field, expressed by   Eq. (6.11), the cartesian components ϵ_{13}, ϵ_{23} et ϵ_{33}, as well as the cylindrical components $\epsilon_{rz}, \epsilon_{\theta z}, \epsilon_{zz}$.*are* all equal to zero Table 6.1 shows the expression of the non-zero cartesian and cylindrical components $\epsilon_{11}, \epsilon_{22}$,and ϵ_{12}, $\epsilon_{rr}, \epsilon_{r\theta}$,and $\epsilon_{\theta\theta}$ as well as the form of the corresponding stress components, derived, on the basis of Table 3.1 and Eq. (3.25).

Table 6.1 Strain and stress components induced by the edge dislocation represented in Fig. 6.2. In this table $C = b/[4\pi(1-\nu)]$ and $D = 2\mu C$.

Non zero Cartesian components	Cylindrical components
$\epsilon_{11} = -C\sin\theta(2 - 2\nu + \cos 2\theta)/\mathbf{r}$ $\epsilon_{22} = C\sin\theta(2\nu + \cos 2\theta)/\mathbf{r}$ $\epsilon_{12} = C\cos\theta\cos 2\theta/\mathbf{r}$	$\epsilon_{rr} = -C(1 - 2\nu)\sin\theta/\mathbf{r}$ $\epsilon_{\theta\theta} = -C(1 - 2\nu)\sin\theta/\mathbf{r}$ $\epsilon_{r\theta} = C\cos\theta/\mathbf{r}$
$\sigma_{11} = -D\sin\theta(\cos 2\theta + 2)/\mathbf{r}$ $\sigma_{22} = D\sin\theta\cos 2\theta/\mathbf{r}$ $\sigma_{33} = -2D\nu\sin\theta/\mathbf{r}$ $\sigma_{12} = D\cos\theta\cos 2\theta/\mathbf{r}$	$\sigma_{rr} = -D\sin\theta/\mathbf{r}$ $\sigma_{\theta\theta} = -D\sin\theta/\mathbf{r}$ $\sigma_{zz} = -2D\nu\sin\theta/\mathbf{r}$ $\sigma_{r\theta} = D\cos\theta/\mathbf{r}$

The relative volume expansion $(\delta\Omega/\Omega) = (\epsilon_{11} + \epsilon_{22}) = (\epsilon_{rr} + \epsilon_{\theta\theta})$, is:

$$\frac{\delta\Omega}{\Omega} = \frac{-b(1 - 2\nu)}{2\pi(1 - \nu)}\frac{\sin\theta}{r} = \frac{-b\mu}{\pi(\lambda + 2\mu)}\frac{\sin\theta}{r} \tag{6.15}$$

Consistent with the choice of the reference frame, the "compressed" region $(\delta\Omega/\Omega < 0)$ corresponds to $0 < \theta < \pi$ $(y > 0)$ and the "expanded" region to $y < 0$. All the components of the stress are functions of the

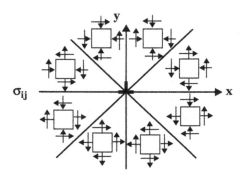

Figure 6.3   Angular variations of the strain components in a plane perpendicular to the edge-dislocation line. The arrows represent the forces acting on the sides of a square within the solid. Their sense expresses the sign of the compression /expansion, or of the shear in each eighth angular fraction of the plane.

angle θ: This stress *is not radial* by contrast to the case of the screw dislocation. Figure 6.3 provides a schematic picture of the angular variations of the stress components in the xy-plane.

Elastic energy

The elastic energy density (Eq. (3.5)) calculated from the components in table 6.1 is :

$$\frac{dU}{d\Omega} = \frac{\mu b^2 (1 - \nu - \nu \cos 2\theta)}{8(1 - \nu)^2 \pi^2 r^2} \qquad (6.16)$$

Hence, the total elastic energy contained between the core of the dislocation (radius r_0) and a cylinder of radius R and of unit height, is:

$$\frac{dU(R)}{dL} = \int_{r_0}^{R} \int_0^{2\pi} \frac{dU}{d\Omega} r dr d\theta = \frac{\mu b^2}{4\pi(1 - \nu)} Ln(\frac{R}{r_0}) \qquad (6.17)$$

This expression is identical, up to the factor $[1/(1 - \nu)]$, to that found for the screw dislocation. This factor being always less than one, an edge-dislocation induces a larger deformation-energy than a screw-dislocation having the same Burgers vector.

6.2.3 *Mixed straight dislocation*

A mixed straight dislocation $\overrightarrow{d}$ parallel to the z axis, is characterized by a Burgers vector having, both, a "screw" component parallel to $\overrightarrow{d}$ and an "edge" component perpendicular to $\overrightarrow{d}$. Since the equilibrium equation 6.1 and the condition expressing the closure defect 6.2 are linear, the displacement $\overrightarrow{\xi}$, the strain $\overline{\overline{\epsilon}}$, and the stress $\overline{\overline{\sigma}}$ induced by this dislocation will be the sum of the corresponding values determined separately by the screw component $(b \cos \alpha)$ and the edge component $(b \sin \alpha)$. Furthermore, the non-zero components of $\overline{\overline{\epsilon}}$ and of $\overline{\overline{\sigma}}$ being distinct for the screw and edge dislocations, expression (3.5) shows that the elastic energy of a mixed dislocation is the sum of the energies calculated for each component, hence, per unit length of the dislocation:

$$\frac{dU}{dL} = \frac{\mu b^2}{4\pi} Ln(\frac{R}{r_0})[\cos^2 \alpha + \frac{\sin^2 \alpha}{(1 - \nu)}] \qquad (6.18)$$

6.2.4 *Elastic energy of a dislocation loop*

The elastic energy induced by a *finite length* of a straight dislocation in an infinite medium is *infinite*. It has been argued hereabove that this property is related to the fact that the formation of a dislocation by slipping requires the application of a shear stress to a portion of plane having an *infinite surface* (unit height and infinite radius). Since the stress decays slowly as a function of the distance, the resulting work is infinite.

In the case of a planar dislocation-loop of the type described in section 5.3.2, we can adapt the preceding argument to justify the *finite* value of the elastic energy induced by such a dislocation. Indeed, the Burgers vector is contained in the plane of the loop, and the surface of slipping is reduced to the surface of the loop. The work of formation of the dislocation will then be finite as well as the elastic energy associated to the dislocation.

An alternate manner of justifying this finite value is to note that a loop necessarily contains pairs of segments having the same Burgers vector but opposite directions $\vec{d}$. The various fields $(6.4)-(6.6)-(6.7)$ calculated for the two members of a pair *considered as isolated* are then of opposite signs. At a sufficiently large distance $(R >> b)$ those fields will compensate eachother to determine a zero-stress. Integrating the energy density, as in Eq. (6.9), can then be performed for a finite radius R, the resulting integral being finite.

It can be shown, more precisely, by general arguments, not reproduced here, that the strain and stress fields induced by a dislocation loop of diameter ρ vanish rapidly beyond a distance $R \approx \rho$, thus justifying the finite character of the elastic energy of the loop.

6.3   Action of a stress on a dislocation

6.3.1   *Effective force applied to a dislocation*

Figure 6.4 recalls the procedure described in chapter 5 (Figs. 5.3 and 5.12) leading to the formation of an edge dislocation by applying a shear stress σ to crystal having finite lengths (A, B, C) in the three dimensions. The position of the dislocation at M within the solid has required the work $W_M = \sigma(Al).b$, in order to produce the translation, by an amount b, of the upper half of the crystal along the surface Al. The dislocation would have been positioned at M' if the work effected by the stress had been $W_{M'} = \sigma(Al').b$. We can consider that the two preceding formations of dislocation at different locations are two successive steps of a given transformation:

a) formation of the dislocation at M;

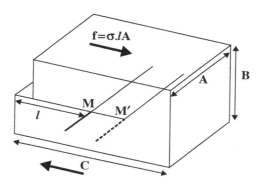

Figure 6.4 Application of a shear stress σ (thick arrows) to an edge-dislocation in order to displace it from M to M' by extension of its gliding surface in the solid.

b) Hypothetical displacement from M to M' by action of the stress σ on the dislocation, associated with the work $(W_{M'} - W_M)$.

The *physical process* of the second step is, in fact, an action on the plane separating the lower and upper halves of the solid to impose a gliding, by an amount, b of an additional length MM'. However, this step can be *interpreted* as the displacement of the dislocation (considered as a physical object) from M to M', submitted to an *effective force* resulting from the presence of the stress σ. If we note F the effective force exerted on a unit length of the dislocation, this force will produce the work $\delta W = F.MM' = F.(l' - l)$. Equalling this quantity to the actual value of the work $\delta W = (W_{M'} - W_M)/A$ produced by the stress, one deduces that $F = b\sigma$.

Beyond this specific situation, it can be shown, more generally, that an effective force, acting on a dislocation, can be defined as a function of σ, and of the local characteristics (direction, Burgers vector) of the dislocation.

6.3.2 *Peach and Koehler formula*

Consider an infinitesimal quasi-straight portion $\overrightarrow{dL}$ of a dislocation D of arbitrary shape (Fig. 6.5). Assume that an internal stress $\overline{\overline{\sigma}}$ present in the volume of the solid surrounding $\overrightarrow{dL}$ has the effect of displacing $\overrightarrow{dL}$ by a translation $\overrightarrow{\delta l}$. The specific situation depicted in Fig. 6.5 is that of a portion of an edge dislocation whose local frame of reference is oriented according to the convention defined in section 5.2.2. The compressed region (containing a portion of an additional half-plane) is situated above the horizontal plane. In order to obtain a displacement of the compressed region in the $\overrightarrow{\delta l}$ direction, the shear force $\overrightarrow{df}$ must be oriented in the sense indicated on the figure, i.e. tending to increase the compression of the region already compressed. The surface of the cut on which this force exerts has the value:

$$\overrightarrow{dS} = \overrightarrow{dL} \wedge \overrightarrow{\delta l} = \overrightarrow{n}.dS \tag{6.19}$$

in which the unit vector $\overrightarrow{n}$, perpendicular to the surface, is directed parallel to the y-axis. On the other hand (cf. Eq. (3.2)), the force $\overrightarrow{df}$ can be deduced from the stress tensor through:

$$\overrightarrow{df} = \overline{\overline{\sigma}}.\overrightarrow{n}\,dS \tag{6.20}$$

As illustrated by the example in the preceding paragraph, the action of the force is to induce a relative slip of the portion of the crystal situated above dS, of amplitude equal to one interatomic distance $\overrightarrow{b}$. Hence, the displacement of $\overrightarrow{df}$ also has the amplitude $\overrightarrow{b}$. The work effected by $\overrightarrow{df}$ is:

$$\delta W = \overrightarrow{b}.\overrightarrow{df} = \overrightarrow{b}.[\overline{\overline{\sigma}}.\overrightarrow{dS}] \tag{6.21}$$

Using the symmetric property of the $\overline{\overline{\sigma}}$ tensor, and also the invariance by permutation of the mixed product of three vectors, this expression can be written:

$$\delta W = (\overrightarrow{b}.\overline{\overline{\sigma}})\overrightarrow{dS} = (\overrightarrow{b}.\overline{\overline{\sigma}})(\overrightarrow{dL} \wedge \overrightarrow{\delta l}) = (\overrightarrow{b}\,\overline{\overline{\sigma}} \wedge \overrightarrow{dL}).\overrightarrow{\delta l} \tag{6.22}$$

The last formula can be interpreted as the work associated to a displacement $\overrightarrow{\delta l}$ of the dislocation (considered as a physical object) submitted to an effective force $\overrightarrow{F}$: $\delta W = \overrightarrow{F}.\overrightarrow{\delta l}$. Identifying the two expressions leads to the so-called *Peach-Koehler* formula, which provides the expression of the effective force.

$$\overrightarrow{F} = \overrightarrow{b}.\overline{\overline{\sigma}} \wedge \overrightarrow{dL} \tag{6.23}$$

The same formula can be established for a portion of a screw dislocation, with the vectors orientation convention defined in section 5.3.1. In consequence, the formula is valid for a portion of any dislocation.

Note that the above formula (B.83) implies that *whatever the type of dislocation and the nature of the local stress* $\overline{\overline{\sigma}}$, *the effective force is perpendicular to the local direction of the dislocation line.*

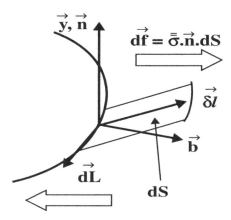

Figure 6.5   Work done by the stress during the displacement of an element of a dislocation dL. The two solid arrows show the direction of the shear stress applied.

The Peach-Koehler formula is valid for any choice of the sense of $\overrightarrow{dL}$ since an opposite choice would also invert the sign of $\overrightarrow{b}$. However, for an edge dislocation, this formula is only valid when the convention defining the sign of $\overrightarrow{b}$ in section 5.2.2 is adopted. An opposite convention would imply to use a different formula, namely: $\overrightarrow{F} = \overrightarrow{dL} \wedge \overrightarrow{b}.\overline{\overline{\sigma}}$.

This important relationship allows to determine the forces exerted on a dislocation of arbitrary shape in a variety of situations: effect of stresses applied to a solid body containing dislocations, interaction between dislocations (section 6.5), interaction between a dislocation and vacancies (section 6.6), etc...

6.4   Line tension of a dislocation

Consider a non-straight portion of a dislocation line (Fig. 6.6 (a)). Its length exceeds that of a straight segment by an amount ΔL. This dislocation induces in the solid a stress field, as well as an elastic energy which can, in principle, be calculated on the basis of the elastic equilibrium equation. For simple shapes of deviations from the straight line, the calculation can sometimes be achieved. It shows that there is an excess of elastic energy with respect to the case of the straight line:

$$\Delta U = g(\Delta L).\mu b^2 > 0 \tag{6.24}$$

in which $g(\Delta L)$ is a positive function which increases with the elongation ΔL. As a consequence, the local deviation will have a tendency to regress spontaneously and restore a straight line. In other terms, a dislocation possesses a certain "stiffness" which opposes deformations.

Such a situation is similar to that of an elastic string which can be deformed by an external force, but which regains its straight shape when the force is suppressed. This property of the string is expressed by its *tension T*, a positive quantity defined by:

$$T = \frac{\partial U}{\partial L} \tag{6.25}$$

in which U is the elastic energy of the string. T is a force directed along the string.

Similarly, a *line tension T* of a dislocation can be defined by the same equation, provided that U is taken as the elastic energy induced in the solid by the dislocation. We can assume that, expanded to lowest order as function of ΔL, $g(\Delta L)$ is equal to $\mathcal{A}\Delta L$ (with $\mathcal{A} > 0$) (note that this expression is exact in the case of a straight dislocation (cf. Eq. (6.18)). We can then write:

$$\Delta U = \mathcal{A}\Delta L.\mu b^2 \tag{6.26}$$

which determines a line tension:

$$T = \mathcal{A}\mu b^2 \tag{6.27}$$

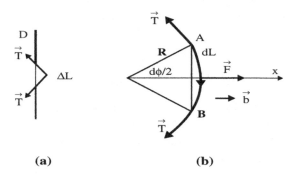

(a)   (b)

Figure 6.6   (a) Straight dislocation possessing a deformed segment. (b) Deformation of a dislocation by a shear stress. The z-axis is towards the front of the figure and the edge-dislocation is oriented in agreement with the convention defined in chapter 5.

The $\mathcal{A}$ coefficient can depend of the global shape of the dislocation, and of its local edge-, screw-, or mixed-, type. Thus for a globally straight dislocation, respectively of the edge- or screw-, type, we derive from Eqs. (6.17) and (6.10):

$$\mathcal{A}_{edge} = \frac{1}{4\pi(1-\nu)} Ln(\frac{R}{r_0}) \qquad \mathcal{A}_{screw} = \frac{1}{4\pi} Ln(\frac{R}{r_0}) \qquad (6.28)$$

Actually, a more precise analysis shows that the proportionnality $\Delta U \propto \Delta L$, to lowest order, is not always valid, and that ΔU can, for instance, be proportional to the logarithm of ΔL. This difference with the situation of an elastic string is related to the fact that a dislocation line exerts a long-range force (slowly decaying with the distance). Hence, the deformation of a portion of a dislocation will interact with the stress field of the entire dislocation, while, for a string, a segment only interacts with the adjacent portions of the string. Nevertheless, quantitative studies of various shapes of dislocations support that the proportionality to ΔL, will, in general, provide consistent results.

As an illustration of this concept, consider a dislocation consisting in a small rectilinear segment AB (6.6b), whose ends are assumed to be "pinned" (i.e. cannot be displaced), by other defects[5] This situation will be exploited in section 8.4.2.

AB is, for instance, an edge dislocation whose Burgers vector $\overrightarrow{b}$ is directed along the x-axis. The stress $\overline{\overline{\sigma}}$ is taken as a pure shear $\sigma_{13} = \sigma > 0$, whose effect is to tend to "close" the angle between the x and z axes. The application of the Peach-Koehler formula (B.83) yields:

$$\overrightarrow{F} = \overrightarrow{b} \begin{bmatrix} 0 & 0 & \sigma \\ 0 & 0 & 0 \\ \sigma & 0 & 0 \end{bmatrix} \wedge \overrightarrow{dL} = \sigma dL \overrightarrow{b} \qquad (6.29)$$

This force acts within the glide-plane $(\overrightarrow{dL}, \overrightarrow{b})$ of the dislocation, and has the effect of inducing the displacement of the portions of the segment which are not blocked in A and B. This displacement will therefore tend to induce a curvature of the segment. *Such a deformation is opposed by the line tension T*. An equilibrium curvature exists, determined by the balance between the effects of σ and of T (Fig. 6.6 (b)). The corresponding radius $R = (dL/d\phi)$ is obtained by projecting the two forces on the $\overrightarrow{x}$-axis. Thus: $F = \sigma b dL = 2T(d\phi/2)$, or, using expression (6.27):

$$R = \frac{T}{b\sigma} = \frac{\mathcal{A}\mu b}{\sigma} \qquad (6.30)$$

---

[5] A possible mechanism of such a blocking will be examined in chapter 8.

A similar expression could be obtained for any type of dislocation since the force $\overrightarrow{F}$ is always perpendicular to the dislocation (cf.Eq. (6.23)). Since a dislocation can only be displaced in its glide-plane (Cf. chapter 8), it is the component F_G of the force in the glide-plane which will determine the curvature. For an applied stress whose value is close to the yield stress of the solid, we have $\mathcal{A} \approx 1$ and $\sigma \approx 10^{-2}\mu$. The resulting curvature corresponds to $R \approx 10^2 b$.

6.5   Interaction between dislocations

In the Peach-Koehler formula (6.23), the stress tensor $\overline{\overline{\sigma}}$ expresses the *local stress* present in the solid surrounding the segment of dislocation considered. If this local stress is due to the presence of a second dislocation in the solid, the resulting force will represent the action of the second dislocation on an element of the first one. The Peach-Koehler formula therefore provides a mean to determine the interaction between two dislocations, through the prior determination of the stress fields induced by each dislocation.[6] To illustrate this determination, let us consider in detail the interaction between two straight and parallel dislocations of various types.

6.5.1   *Qualitative study*

Examine first the *energy and geometrical* considerations which lead to a *qualitative* prediction of the effect of the interaction between dislocations.

Two edge-dislocations in the same glide plane

Consider first two parallel edge-dislocations, located in the same glide-plane Dx, and having the same Burgers vector $\overrightarrow{b}$ (Fig. 6.7 (a)). The compressed regions (containing an aditional half lattice plane) relative to the two dislocations are situated on the same side of the glide-plane. Bringing closer the two dislocations would have the effect of increasing the deformation, and accordingly, the energy of the medium in the interval separating them. They will therefore have a tendency to *repell* eachother.

The same conclusion is reached when the total elastic energy is evaluated as a function of the distance between the dislocations (Fig. 6.7). If this

---

[6]In the simplified approach generally adopted, the stress field of each dislocation is assumed to be the one determined for the dislocation being islolated in the solid. Clearly the presence of two dislocations close to each other will change the stress fields.

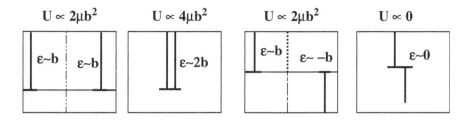

Figure 6.7 (Left) Interaction of two edge dislocations possessing the same glide-plane (x) and the same Burgers vector. The compressed regions are above x. (Right) Edge dislocations with opposite Burgers vectors. Bringing together the two dislocations will restore a complete plane with the two half planes of the two dislocations: no dislocation is eventually left.

distance, R is very large ($R >> b$), the stress field of each dislocation, which decays as $1/r$ can be considered, approximately, to possess significant values in a radius $R/2$. The stress fields of the two dislocations will then concern spatially distinct volumes of the solid. In the framework of this approximation, the total elastic energy will be the sum of their respective elastic energy, proportional to the square of their Burgers vector: $E \approx 2\mu b^2$.

Assume the two dislocations are close enough to have $R \approx b$ and R much smaller than the dimensions of the sample. In most of the volume of the solid we can consider that the stress field is that of a single dislocation associated to the insertion of *two half-planes*. Such a dislocation has a Burgers vector equal to $2\vec{b}$ (Fig. 6.7). The elastic energy is then $E' \approx \mu(2b)^2 = 4\mu b^2 > E$. In agreement with the preceding inference, it appears that bringing together the two dislocations increases the elastic energy. Hence the two dislocations repel each other.

Likewise, if the two dislocations have opposite Burgers vectors ($\vec{b}_2 = -\vec{b}_1$) (Fig. 6.7), the same type of argument shows that $E' \approx 0 < E = 2\mu b^2$. The two dislocations therefore attract each other. The compressed regions, containing an additional half lattice-plane are situated on either part of the glide plane. Bringing closer the two dislocations has the effect of compensating the compressed zone of one dislocation by the expanded zone of the other, thus decreasing, everywhere in the volume of the solid, the magnitude of the strain.

If the mobility of the two dislocations permits their motion in the solid, they will decrease their distance and their respective half-planes will form a complete lattice plane thus suppressing the presence of dislocations in the crystal. One can say that such two dislocations of opposite sign *cancel mutually*.

Two parallel screw-dislocations

For two parallel screw-dislocations having the same Burgers vector, the energy-argument used for edge-dislocations applies. Two distant dislocations induce an elastic energy $\approx 2\mu b^2$ while two close dislocations possess the energy $\approx 4\mu b^2$. These dislocations therefore repell eachother. Likewise, two screw-dislocations with opposite Burgers vectors attract eachother. One is associated with a "right-handed helix" formed by interconnected planes perpendicular to $\vec{b}$, and the other to a left-handed helix having the same pitch. If they are brought together to coincide, no defect is left: they cancel mutually.

One screw-dislocation and one, parallel, edge-dislocation

Let D_1 and D_2 be respectively a screw dislocation and an edge-one, both parallel to the z-axis, and with Burgers vectors $\vec{b}_1$ (parallel to z) and $\vec{b}_2$ (perpendicular to z). The effective force exerted on a dislocation has its origin in the action of the local stress on the cut-plane which has enabled to form or displace the dislocation. The plane defined by D_1 and D_2 can be taken (cf. Fig. 6.8) as the cut-plane common to the two dislocation (such a plane containing D_1 being arbitrary in the case of a screw dislocation).

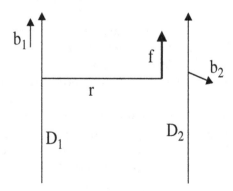

Figure 6.8   Interaction of a screw dislocation with a parallel edge-dislocation.

The force $\vec{f}_{12} = \overline{\overline{\sigma}}_1 . \vec{n} = \overline{\overline{\sigma}}_1 . \vec{y}$ exerted by D_1 on the plane $D_1 D_2$ is *parallel to* z since a screw dislocation induces in the solid medium a pure shear stress $\sigma_{\theta z}$ (Eq. (6.7)). The work produced by such a force for a displacement parallel to $\vec{b}_2$ is therefore equal to zero. Hence, D_1 and D_2 *do not interact.*

6.5.2 Quantitative study

Forces between parallel screw dislocations

D_1 and D_2 are two parallel screw dislocations oriented along $z > 0$ with Burgers vectors $\vec{b}_1$ and $\vec{b}_2$ distant by R.

The stress field of a screw dislocation has cylindrical symmetry, the origin of the z-axis being taken on D_1. The x-axis can therefore be chosen in the plane $D_1 D_2$ of the two dislocations. The only non-zero component of the stress generated by D_1 and acting on D_2 is $\sigma_{23} = (\mu b / 2\pi R)$. The application of the Peach-Kohler formula to the unit-length of D_2 yields:

$$\vec{F}_{12} = \vec{b}_2 . (\frac{\mu b_1}{2\pi R}) \begin{bmatrix} 0\,0\,0 \\ 0\,0\,1 \\ 0\,1\,0 \end{bmatrix} \wedge \left(\begin{bmatrix} 0 \\ 0 \\ 1 \end{bmatrix} \right) = \frac{\mu b_1 b_2}{2\pi R^2} \vec{R}_{12} \qquad (6.31)$$

This force is directed along the radius, such as $\vec{F}_{12} = -\vec{F}_{21}$, *repulsive* if the two dislocations have the same sign (two right-handed screw-dislocations or two left-handed ones) and *attractive* if they are of opposite sign. Its modulus is $|F| = \mu |b_1 b_2| / 2\pi R$.

Forces between parallel edge-dislocations

Let D_1 and D_2 be two straight edge-dislocations, both parallel to z and whose Burgers vectors are parallel to x (Fig. 6.2).[7] Their glide planes $(\vec{d}, \vec{b}_i)$ are parallel. Using the cartesian coordinates of the stress tensor generated by D_1 (table 6.1), and the Peach-Koehler formula, one finds that the force exerted on D_2 is:

$$\vec{F}_{12} = \frac{\mu b_1 b_2}{2\pi R(1-\nu)} \begin{pmatrix} \cos\theta \cos 2\theta \\ \sin\theta(\cos 2\theta + 2) \\ 0 \end{pmatrix} \qquad (6.32)$$

---

[7]The case of Burgers vectors which are not parallel will be considered in exercise 15 (their glide planes then intersect each other).

Figure 6.9 shows, for two dislocations having the same sign, the direction and sense of the x and y components of the forces acting on D_2. The sense would be inverted for dislocations of opposite signs.

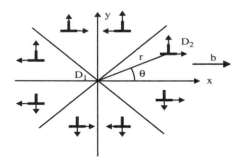

Figure 6.9   Interaction between two edge-dislocations parallel to z, with Burgers vectors parallel to x.

As will be argued in chapter 8, unless a material is heated at high-enough temperatures, D_2 can only be displaced by forces F_x parallel to the glide-plane ($\vec{d}$, $\vec{b}_i$). In such a situation, two dislocations having the same sign will repell each other if $x^2 > y^2(-\pi/4 < \theta < \pi/4)$ or if $3\pi/4 < \theta < 5\pi/4$). Moreover, they will tend to *move above each other on the y axis if* $x^2 < y^2$. The axes $x = \pm y$ are metastable locations for the dislocation D_2. Similar conclusions can be drawn for dislocations of opposite sign.

If both the components F_x and F_y can displace the dislocation D_2, a situation which can arise at high temperatures (cf. section 8.8) The result of the interaction is the same as for two screw dislocations: repulsion for dislocations of same sign and attraction for dislocations of opposite sign.

Parallel dislocations of mixed types

The stress fields induced in a solid by a straight mixed-type dislocation is the sum of the fields corresponding to the screw component and to the edge-component (cf. section 6.2.3). Since a screw component does not interact with an edge component, the resulting interaction will derive, on one hand from the interaction between the two screw components, and on the other hand from the interaction between the two edge-components. The respective forces are determined by the Eqs. (6.31) and (6.32).

Non-parallel dislocations

If two dislocations are not parallel, their interaction is no longer invariant in the z-direction of one of the dislocations. The force produced by D_1 on each element of D_2 is not uniform. The global action on D_2 will generally comprise a set of forces, with various orientations, distributed along D_2. A number of effects can result: attraction or repulsion between dislocations, rotation, and bending of each dislocation, formation of junctions or of "jogs" (see chapter 8).

6.6 Interaction between a dislocation and a vacancy

As shown in chapter 4, a vacancy induces at its site a local volume expansion $\delta\Omega$. In turn, such an expansion generates strain and stress fields. In general, such fields will exert a force on a dislocation present in the solid, in agreement with the Peach-Koehler formula. The opposite force acting on the vacancy and produced by the dislocation stress field, is easier to calculate (Fig. 6.10). Indeed, if a continuous medium is submitted to a stress field σ_{ij}, the elastic energy required to create a vacancy in it (cf. section 4.3.1), by withdrawing from the medium a sphere of matter and replacing the void by a rigid sphere of smaller or larger size, is $-p\delta\Omega$ where $p = -(1/3)\sum \sigma_{kk}$ is the hydrostatic pressure corresponding to the stress field.

Since a screw dislocation does not induce a local pressure (cf. Eq. (6.7)), there is no interaction of such a dislocation with a vacancy (or any other spherical point defect).[8]

In the case of an edge dislocation the pressure exerted at a distance R (cf. Table 6.1) determines the following energy transmitted to the point defect:

$$U = -p\delta\Omega = \frac{-2D(1+\nu)\delta\Omega \sin\theta}{3R} = \frac{-\mu b(1+\nu)\delta\Omega \sin\theta}{3\pi(1-\nu)R} \qquad (6.33)$$

---

[8]This conclusion fails if the considered solid medium is not isotropic. Besides, certain point defects, as interstitials can induce in a solid a shear component of deformation. In both cases there will be an interaction between a screw dislocation and the point defect.

A radial force $F = (-\partial U / \partial R)$ is derived from it.

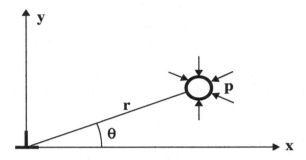

Figure 6.10   Interaction between an edge dislocation and a point defect. The interaction puts into play the local pressure induced by the dislocation.

Chapter 7

Interactions with the lattice

Main concepts: Nature of the interactions which determine the width of the dislocation core. Influence of the type of binding of the crystal (covalent, metallic, etc...). Peierls-Nabarro pinning force. Dissociation of a dislocation core.

7.1 Introduction

This chapter is devoted to the study of the interaction between the dislocation core and the microscopic structure of a crystalline solid. It is this interaction which accounts for the occurence of *fragile materials*, which break without prior deformation when a stress of large-enough magnitude is applied to them. It also allows to understand the occurence of *ductile materials* which can be deformed plastically, by a more or less important amount, before breaking. In substances which possess a small ductility, this interaction is predominant in determining the *yield strength* (the elastic limit), especially in non-metallic solids. Finally, its characteristics explain that the "easy gliding" planes of a dislocation have specific crystallographic directions.

The study of this interaction requires taking into account the details of the crystal structure of a material. It involves a schematic atomic model of the dislocation core.

In this chapter, attention is first given to the atomic configuration in this model. The present knowledge of the structure of the core relies essentially on the results of numerical models, specific of given materials and of certain types of dislocations. A few high-resolution electron microscope observations are also available. Hence, this knowledge is largely speculative, and it is legitimate to discuss here a schematic microscopic model, which, although very simplified, has the advantage, as will be shown in chapter

8, to account qualitatively, in a satisfactory manner, for the diversity of the plastic behaviour in existing materials. This model emphasizes the fact that the properties of the dislocation core (and, in particular, its width) depend on the relative importance of two characteristic features of a crystal structure: a) Its elastic stiffness in the direction parallel to the Burgers vectors, and, b) The more or less pronounced anisotropy of the chemical bonds linking atoms across the glide plane of the dislocation.

In a second step, it is emphasized that there exists a minimum, suitably oriented, stress (called the Peierls-Nabarro stress) which has to be applied to a dislocation in order to "unpin" it from the crystal lattice, and displace it in a direction parallel to the Burgers vector.

In the last paragraph a more complex feature of the dislocation core in certain materials is discussed, namely, the local dissociation of the dislocation into two "partial" (or "imperfect") dislocations.

7.2   Core structure of an edge dislocation

In chapter 5 the core of a dislocation was defined as the *region of transition* between two parts of a crystal which have slipped with repect to eachother, and in which the atomic configuration is strongly perturbed, as compared to the ideal crystal. Figures 7.1 and 7.2 decompose the mechanism of formation of an edge-dislocation with the view of specifying a schematic geometry of the core. A standard choice of reference axes is made: x parallel to $\vec{b}$, y pointing to the additional half-plane, z parallel to the dislocation. The lattice periods are assumed to be b along both x and z, and a along y. The glide plane of the dislocation is the xz plane.

The first step of the hypothetical procedure of formation is represented on the left part of Fig. 7.1. Its consists in performing a cut along the left horizontal half-plane, and imposing a slipping of the upper part of the crystal by an amplitude b. This partial slipping, which encounters, as a rigid obstacle, the right part of the crystal, determines the occurence of an additional half lattice plane.

The centre of Fig. 7.1 represents the symmetrical configuration of deformed lattice planes, which is reached when the forces imposing the slipping have been suppressed, and when the chemical bonds across the cut have been restored.

Finally, the right part of Fig. 7.1, shows schematically that the same symmetric state can be obtained by performing a cut along *the entire glide*

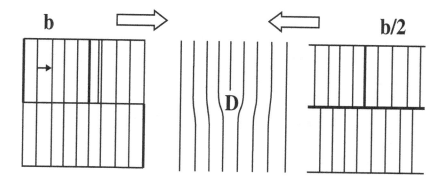

Figure 7.1   Steps of formation of an edge dislocation. (Left) Partial slipping of the left upper part of the crystal. (Center) Symmetrization of the configuration of lattice planes. (Right) Symmetric configuration obtained by a uniform sliding of $(b/2)$.

plane and in translating by $(b/2)$ the upper half crystal. The configuration thus obtained is already symmetric. Restoring the bonds across the cut plane will result, again, in establishing the configuration represented in the center. This second procedure will underly the mathematical description of the core structure.

In Fig. 7.2 a more detailed picture of the final stage of the formation mechanism is shown, with neighbouring atoms, and the configuration of the bonds between atoms across the glide-plane $(y = 0)$. These bonds will favour a restoration of the equilibrium configuration in the ideal crystal, i.e. relative positions of the atoms they link on the same vertical line. Hence, they will tend to "push" the vertical half lattice planes situated *above* the glide plane $(y > 0)$, closer to the center of the figure (where the additional half-plane is located). Conversely, it tends to "pull" the vertical half lattice planes situated *below* the glide plane $(y < 0)$ further from the center of the figure. From these displacements an x-dependent shift $\xi(x)$ of the position of the different vertical half lattice planes will result.

7.2.1   *Lattice planes coordinates*

The origin O of the x-axis is taken at the center, on the additional half lattice plane of the crystal. Other vertical half lattice planes above the glide plane (or alternately, the atoms situated at their lower tip) are located at x_n, while half-planes below the glide plane have the coordinate x'_n. Prior to the restoration of the bonds across the glide plane (right part of Fig. 7.1), these coordinates are:

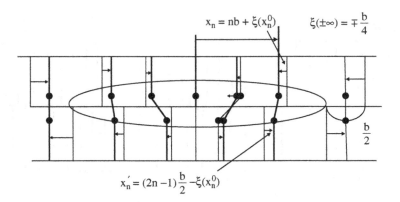

Figure 7.2   Geometrical configuration of the core of a dislocation in the neighbourhood of the additional half-plane.

$$y > 0 \qquad x_n^0 = nb \ (n = \pm 1, \pm 2, ...) \tag{7.1}$$

$$y < 0 \qquad x_n'^0 = (2n - 1)\frac{b}{2} \ (n > 0) \ ; \ x_n'^0 = (2n + 1)\frac{b}{2} \ (n < 0) \tag{7.2}$$

After restoration of the bonds, and "relaxation" of the atom positions, (Fig. 7.2), there is a shift of each half-plane with respect to the above positions (Eq. (7.1)). The set of shifts $\xi(x_n^0)$, $\xi(x_n'^0)$ forms a displacement field, assumed to be parallel to x, and *whose equilibrium value is yet undetermined*. Note that atoms facing eachother across the glide-plane will have opposite shifts, since these shifts are induced by a common bond : $\xi(x_n^0) = -\xi(x_n'^0)$. In the "compressed" region $(y > 0)$, the atomic positions are therefore:

$$y > 0 \qquad x_n = nb + \xi(x_n^0) \ (n = \pm 1, \pm 2, ...) \tag{7.3}$$

with $\xi < 0$ at the right of the dislocation and $\xi > 0$ at its left. In the "expanded" region $(y < 0)$:

$$y < 0 \qquad x_n' = (2n - 1)\frac{b}{2} - \xi(x_n^0) \ (n = 1, 2, ..) \tag{7.4}$$

$$y < 0 \qquad x_n' = (2n + 1)\frac{b}{2} - \xi(x_n^0) \ (n = -1, -2, ...) \tag{7.5}$$

Obviously, the bond distortion imposed by the dislocation will decrease as a function of the distance to D along x. Hence for $x_n^0 = \pm\infty$, an ideal structure exists, with the half planes on either side of the glide plane, forming a complete vertical lattice plane: $x_n'(+\infty) = x_n(+\infty)$ and $x_n'(-\infty) = x_n(-\infty)$. One then deduces from Eq. (7.1), that $\xi(+\infty) = -(b/4)$ and $\xi(-\infty) = (b/4)$.

7.2.2 *Components of the core energy*

The atomic configuration in the core, being distinct from the equilibrium configuration in the ideal structure, determines an excess of energy with respect to an identical volume in the perfect crystal. Two components of this excess-energy can be distinguished.

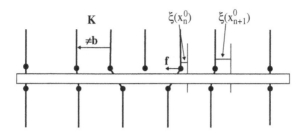

Figure 7.3   Excess elastic energy, related to the variations of the lattice plane distances on either side of the glide plane.

i) An "elastic" component, which is related to the fact that, on both sides of the glide plane (in the upper and lower halves of the crystal), the (non-uniform) distances between vertical lattice planes (along x) deviate from their equilibrium distance in the ideal crystal.

ii) A "misfit" energy, consequence of the non-alignment of atoms facing eachother, on either sides of the glide plane, on the same vertical plane. This excess energy is due to the disorientation of the bonds linking these atoms.[1]

Elastic component of the core excess-energy

The equilibrium interplane distance parallel to x is b (Fig. 7.3). For *small deviations* with respect to this distance, the elastic energy can be written as:

$$U_{elast} = 2 \sum_{n} \frac{K}{2} [\xi(x_{n+1}^0) - \xi(x_n^0)]^2 \qquad (7.6)$$

---

[1]The standard theory developed in most works on the subject, express this second component by introducing an infinite, continuous, distribution of dislocations possessing infinitesimal Burgers vectors. The resulting equation is very similar to the one obtained here, but its physical interpretation is less clearly related to the characteristics of the bonds between atoms.

with K positive. The factor of 2 expresses the fact that the upper and lower part of the crystal contribute equally to this energy.

From this expression, it can be infered that the atom (or the corresponding lattice plane) located at x_n is submitted to a restoring force equal to:

$$f_{elast} = K[\xi(x_{n-1}^0) - \xi(x_n^0)] - K[\xi(x_n^0) - \xi(x_{n+1}^0)] \tag{7.7}$$

In the compressed region ($y > 0$), this force is directed to the right ($f_{elast} > 0$), for $x_n > 0$ and to the left for $x_n < 0$. The directions are reversed in the expanded half of the crystal ($y < 0$).

A *qualitative* comparison of the values of this energy for a "narrow" core (having a width $W \sim b$), and for a "wide" core (width $W >> b$) is of interest. The total variation of ξ, namely $|\Delta\xi| = (b/2)$ is the same in both cases. Assume that this variation is linear accross the width. For a narrow core, of width $\sim b$, the above formula yields $U_{elast} \approx Kb^2/4$. For a wide core, containing many lattice planes, putting $N = (W/b)$, the formula results in:

$$U_{elast} = KN \left\{ \frac{b}{2N} \right\}^2 = (\frac{Kb^2}{4})(\frac{b}{W}) << \frac{Kb^2}{4} \tag{7.8}$$

Hence, *a width of the core as large as possible will correspond to the smallest elastic energy.*

Misfit across the glide plane

The two atoms with coordinates x_n and x_n', situated on either side of the glide plane, are shifted by $\Delta(x_n) = (x_n - x_n')$ in the direction parallel to x. This shift induces a restoring force tending to cancel Δx_n in order to have a bond perpendicular to the glide-plane as in the ideal crystal (Fig. 7.4).

To evaluate the magnitude of the restoring force, a simplified argument can be used. Thus, the rotation of a bond can be obtained by imposing a slipping of the upper part of the crystal with respect to the lower part, through application of a shear stress $\sigma_{xy} \propto \mu$ (Fig. 7.4 (b)).

Note that the required stress $\sigma_{xy} = \sigma$ is a periodic function of $\Delta(x_n)$ of period b. Indeed, a slipping of amplitude b, determines an atomic configuration identical to the initial configuration (by substitution of an atom by the neighbouring one). Hence, $\sigma[\Delta(x_n) + b] = \sigma[\Delta x_n]$. The simplest expression of such a function, which vanishes for $\Delta(x_n) = 0$, is :

$$\sigma = \pm A \sin \frac{2\pi \Delta(x_n)}{b} \tag{7.9}$$

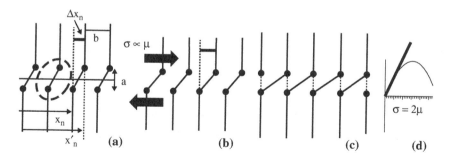

Figure 7.4   (a)−(c) Evolution of the shear across the glide plane, obtained by a gradual increase of the stress σ. The disorientation of the bonds across the glide plane generate an increase of energy. In (c), up to a primitive translation, there is no misorientation and the excess energy is equal to zero. (d) Shows that the shear stress is a periodic function. It also shows the value of the slope of σ at the origin.

Its sign depends of the region considered (the compressed one or the expanded one). The value of A can be deduced from the following remark: for a very small displacement $\Delta(x_n)$ the relationship between the shear $\epsilon = [\Delta(x_n)/2a]$ and the stress must be that corresponding to the elastic behaviour, i.e. to Hooke's law, namely: $\sigma_{xy} = 2\mu\epsilon$. Linearizing (7.9), yields $2\pi A \Delta x/b = \mu \Delta x/a$. We derive from this value of A :

$$\sigma = \pm \frac{\mu b}{2\pi a} \sin \frac{2\pi \Delta(x_n)}{b} \qquad (7.10)$$

On the other hand, the shift $\Delta(x_n)$ between two atoms (x_n, x'_n) on either sides of the glide plane, can also be expressed as a function of the displacement ξ in (Eq. (7.3)):

$$\Delta(x_n) = x_n - x'_n = 2\xi(x_n^0) \pm \frac{b}{2} \qquad (7.11)$$

Finally:

$$\sigma = \mp \frac{\mu b}{2\pi a} \sin \frac{4\pi \xi(x_n^0)}{b} \qquad (7.12)$$

The maximum value of the stress, $\sigma_{\max} = (\mu b/2\pi a)$, called the *upper limit of the yield stress* is equal to the stress needed to form a first dislocation in a crystal (see chapter 5, note 2). This value will be discussed again in section 8.5.1.

7.2.3   Determination of the core width

The atom located at x_n is in equilibrium when the two opposite forces f_{elast} (Eq. (7.7)) and $f_{misfit} = \sigma b^2$ (Eq. (7.12)) have the same magnitude:

$$K[\xi(x_{n-1}^0) - 2\xi(x_n^0) + \xi(x_{n+1}^0)] = -\frac{\mu b^3}{2\pi a} \sin \frac{4\pi\xi(x_n^0)}{b} \tag{7.13}$$

This equation which determines the unknown function ξ of a discrete variable x_n^0, a coordinate which depends of the integer index n. It can only be solved numerically. However, to obtain some insight on the characteristics of this function, assume for the sake of simplicity, that the core is wide enough ($W \gg b$) to consider that, at the scale of W, the x_n^0 as a quasi-continuum set of values, close to eachother. Equation (7.13) can then be transformed into a differential equation bearing on the function $\xi(x)$. In this view, one can write, since $(x_{n+1}^0 - x_n^0) = (x_n^0 - x_{n-1}^0) = b$:

$$\xi(x_{n-1}^0) - 2\xi(x_n^0) + \xi(x_{n+1}^0) = b^2 \frac{d^2\xi}{dx^2} \tag{7.14}$$

Equation (7.13) then becomes:

$$\frac{d^2\xi}{dx^2} = -\frac{\mu b}{2\pi a K} \sin \frac{4\pi\xi(x)}{b} \tag{7.15}$$

The solution must comply with the limit conditions $\xi(\pm\infty) = \mp b/4$ and $\xi'(\pm\infty) = 0$. Multiplying the two members of the equation by $(d\xi/dx)$, integrating, and taking into account the limit conditions, one obtains:

$$\frac{d(\pi\xi/b)}{dx} = \pm\sqrt{\frac{\mu}{2aK}} \cos(2\pi\xi/b) \tag{7.16}$$

If $t = \tan(\pi\xi/b)$, and $(W/2) = \sqrt{2aK/\mu}$, a quantity which has the dimension of a length, the following solution $\xi(x)$ is found, which complies with the imposed limit conditions $\xi(\pm\infty) = \mp(b/4)$:

$$\xi(x) = -\frac{b}{\pi} \arctan[th\frac{2x}{W}] \tag{7.17}$$

The atomic shift $\xi(x)$ is represented on the figure hereunder:

For $x = \pm W/2$, the value of ξ is $\approx 0.2b$. The quantity W can be considered as the *width* of the core. It is in this interval that $\xi(x)$ has the essential part of its variation. In agreement with the qualitative argument developed in the previous paragraph, it appears that the occurrence of a "wide" core is favoured by a large value of the "longitudinal" elastic stiffness K as well as by a small shear stress (small value of μ).

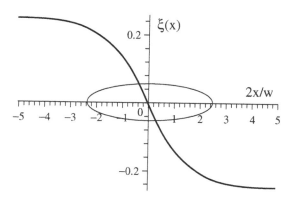

Figure 7.5   Variations of the shift $\xi(x)$ between atoms across the glide plane, within the core of an edge dislocation.

A numerical evaluation of W is of interest. From its definition in (7.7), the coefficient K can be infered to be of the order of Ca, in which C is a stiffness constant (cf. chapters 1 and 3). Hence, $W \approx 2a\sqrt{2C/\mu}$. Referring to Table 3.3, it appears that, e.g. for metals having an FCC structure, $2C/\mu \approx 10$, henceforth $W \approx 5a \approx 5b$. Such a result only provides us with an order of magnitude, since the model considered is schematic.

Various other models of a dislocation core have been studied which differ, either by the form of the compression elastic stress (7.7) or by that of the shear stress (7.12). All these models determine a core-width of the order of a few interatomic distances (i.e. a few times b).

Such a result ($W \approx b$) could cast a doubt on the validity of the "continuous approximation" used, relying on the assumption $W >> b$. However, numerical studies of the original discrete Eq. (7.13), show that, in spite of this inconsistency, the results obtained here are qualitatively correct.

The expression obtained for W shows that the easier the shear of the crystal along the glide plane, the larger the width of the core. This will occur if the chemical bonds across the glide plane are "tolerant" to a variation of the angle between bonds. From this standpoint, it appears that a "covalent" solid will be less tolerant than a metal. Indeed the orientation of a covalent bond is very precisely defined, a disorientation being associated to a steep increase of energy (cf. Fig. 7.6). By contrast, in a metal (cf. chapters 2 and 4), the cohesion is realized by the electron gas which establishes almost *isotropic* bonds between neighbouring atoms.

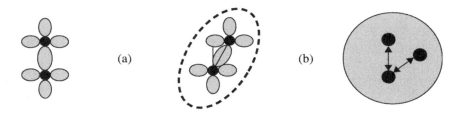

Figure 7.6   Angular dependence of the bond between atoms (a) in a covalent solid, and (b) in a metal in which the electron gas determines a quasi-isotropic bond.

On the other hand, for a given relative shift $\xi(x_n^0)$, the shear is smaller if the ratio (a/b) is larger. Note that, in the crystal structure considered, the maximum value of this ratio is realized if the glide plane coincides with one of the most dense lattice planes (cf. chapter 2). Hence, the width of a dislocation-core will be larger if its glide plane is a dense lattice plane.

7.3   Peierls-Nabarro stress

In this section, a wide dislocation core is shown to determine an easier motion of the dislocation in the solid.

7.3.1   *Gliding of a dislocation*

The edge-dislocation represented on Fig. 7.7 (a) is in a stable configuration. Indeed, it is symmetric with respect to the additional half plane located in the upper part of the crystal. The shearing forces, of the form $[\propto \sin 4\pi\xi(x_n^0)/b]$, which act on the atoms x_n at the right of the symmetry plane, and which would tend to "push" the dislocation to the left, are exactly balanced by the homologous forces acting on the atoms situated at the left of the symmetry plane.

This configuration is not the only stable one. Figure 7.7 shows that if the upper part of the crystal is translated by one interatomic spacing b parallel to x, an identical symmterical configuration is realized. By contrast, a translation by an amplitude $\alpha.b$ $(0 < \alpha < 1/2)$ will establish an asymmetric configuration. The sum of the forces exerted on the atoms at the right and at the left of the additional half-plane will be unbalanced, and they will tend to displace the dislocation towards its nearest symmetric configuration.

Thus, the displacement of a dislocation between two consecutive stable positions, encounters opposing forces. The displacement must overcome an *energy barrier* determined by the shearing forces. The potential energy $U_{cis}(x)$ associated to these forces (Fig. 7.8) thus possesses a succession of minima of equal depth separated by b. Each minimum is called a *Peierls valley* : parallel to the direction z of the dislocation, the energy is constant. These minima are separated by maxima (the energy barrier) located at mid-distance between the stable positions on the x -axis, and which are unstable, although symmetrical, positions of the dislocation.

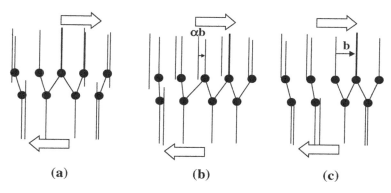

(a)   **(b)**   **(c)**

Figure 7.7   Effect of the displacement of a dislocation parallel to the Burgers vector. A shift of amplitude b leads to an identical local atomic configuration.

If $U_{cis}(\alpha.b)$ is the energy required to displace the dislocation from a position nb to a neighbouring position $(nb + \alpha b)$, the value of the restoring shear stress acting on the dislocation to bring it back to nb, is $\sigma(\alpha b) = -\partial U_{cis}/\partial(\alpha b)$. The maximum value of this stress, as a function of α, is called the *Peierls-Nabarro stress* σ_{PN}. Applying this stress allows to achieve the displacement of the dislocation, in its glide plane, from one stable position nb to the neighbouring one $((n+1)b$ or $(n-1)b)$.

The *principle* of the computation of σ_{PN} is described in the next section. It will be shown that σ_{PN} vanishes if the width W of the core is very large (the dislocation can then slide freely along x). Its expression will also be given a s a function of W. Its value is, in general, much smaller than that of the *upper limit of the yield stress*, (Eq. (7.12)).

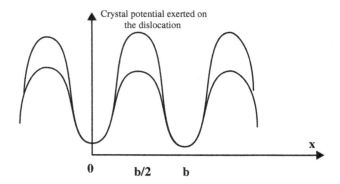

Figure 7.8 Schematic variation of the potential acting on a dislocation parallel to its glide direction. It is exerted by the crystal atomic structure. Its minima are the "Peierls valleys".

7.3.2 *Principle of calculation of the Peierls-Nabarro stress*

The restoring shear stress $\sigma(\alpha b)$ is the sum of the local shears $\sigma(x_n^0 = nb + \alpha b)$ expressed by (7.12) and associated to the relative shift of the atoms of the upper half crystal with respect to their counterpart in the lower half-crystal. Thus:

$$\sigma = \frac{\mu b}{2\pi a} \sum_{n=-\infty}^{n=+\infty} \sin \frac{4\pi \xi(x_n^0)}{b} \tag{7.18}$$

Although the function $\xi(x)$, expressed by Eq. (7.17), has been determined in the framework of the quasi-continuous approximation considered here-above, it will be assumed that the function (7.17) describes correctly the atomic displacements $\xi(x_n^0)$, even when these atomic positions have discrete values. The form of $\sigma(x_n^0)$ is then determined by the set of two equations (7.18) and (7.17).

When $x_n^0 = nb$, corresponding to the stable configuration of the dislocation, located in the bottom of the Peierls valleys, the stress σ is necessarily equal to zero, because the terms (n) and $(-n)$ in (7.18) have opposite values (symmetry of the dislocation). This conclusion agrees with the qualitative conclusions of the preceding section.

An arbitrary configuration is defined by $x_n^0 = (nb + \alpha.b)$ with $(0 < \alpha < 1)$. If the dislocation-core is very wide $(w >> b)$, the stress σ also equals zero for any translation $\alpha.b$. Indeed, in this case, (7.18) can be written as an integral since the successive x_n^0 form a quasi continuum at the scale of the width of the core. Hence, taking into account the number (dx/b) of lattice planes in an interval $dx \gg b$, and the equality (7.15), the sum 7.18 becomes:

$$\sigma = \frac{\mu}{2\pi a} \int_{-\infty}^{+\infty} \sin \frac{4\pi \xi(x)}{b} dx = \frac{K}{b} \int_{-\infty}^{+\infty} (\frac{d^2\xi}{dx^2}) dx = \frac{K}{b} \left[\frac{d\xi}{dx}\right]_{-\infty}^{+\infty} = 0$$
(7.19)

The physical reason underlying this cancellation is that the shifts $\xi(x_n^0)$ also form a quasi-continuum of values in the interval $(\pm b/4)$. A displacement of the core by an arbitrary value $a.b$ will therefore have the only effect of exchanging the different $\xi(x_n^0)$ values present in the sum (7.18). As a consequence, $\sigma(nb + a.b) = \sigma(nb) = 0$.

The qualitative discussion of the discrete situation (cf. Fig. 7.7) has shown that, for a narrow core, the stress required to displace the dislocation is not equal to zero. The vanishing of this stress for a very wide core (Eq. (7.19)) therefore suggests that the maximum stress σ_{PN} *decreases* when the width of the core increases. The actual calculation of the discrete sum (7.18), will not be reproduced here. This calculation as well as the calculation of similar discrete sums relative to other schematic models of the core, show that the Peierls-Nabarro stress has the form:

$$\sigma_{PN} \approx 2\mu \exp(-\frac{\gamma w}{b})$$
(7.20)

in which γ is a dimensionless coefficient of the order of a few units ($\gamma = 2\pi$ or $\gamma = \pi$ depending of the model considered). For instance, if $\gamma = 2\pi$, this formula determines the value $\sigma_{PN} \approx 4.10^{-3}\mu$ for $w \approx b$, and $\sigma_{PN} \approx 10^{-5}\mu$ for $w \approx 2b$. Those values are considerably less than the value of *upper limit of the yield stress* $(\mu b/2\pi a)$ in Eq. (7.10).

7.4   Dissociation of the dislocation core

Microscopic observations as well as numerical simulations have revealed that the real structure of a dislocation core can be more complex than the model examined in the preceding sections. One of the encountered features, the dissocation into two "partial" dislocations can be discussed on a simple basis.

As indicated in section, the favoured glide plane orientation of a dislocation are the ones with the largest atomic density. In a metal having an FCC structure, the highest density planes are those perpendicular to the

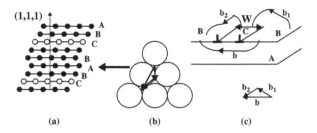

Figure 7.9   Dissociation of a dislocation. The discontinuous displacement b induced by the dislocation is decomposed into two separate discontinuous displacements b_1 and b_2. This decomposition is associated to a stacking fault of width W.

main diagonals of the cubic cell. The atomic structure of these planes is hexagonal compact (cf. Fig. 2.15 in chapter 2). As shown on this figure, they form a stacking of planes, according to a sequence ABCABC.. , in which the atoms of the B planes are shifted with respect to those of the adjacent A planes by $\vec{g} = \frac{1}{6}(\vec{c}_1 - 2\vec{c}_2 + \vec{c}_3)$ parallel to the plane orientation (the $\vec{c}_i$ are the edges of the cubic cell). A C-plane is also shifted by $\vec{g}$ with respect to the adjacent B plane.

Consider a dislocation D formed by a slip (limited to D) of a portion of B-plane, normal to the main diagonal $(1,1,1)$. The possible Burgers vectors of this dislocation are the shortest translations parallel to B. These are the six vectors $\pm\frac{1}{2}(\vec{c}_1 - \vec{c}_2), \pm\frac{1}{2}(\vec{c}_1 - \vec{c}_3), \pm\frac{1}{2}(\vec{c}_2 - \vec{c}_3)$. Let the Burgers vector be, for instance, $\vec{b} = \frac{1}{2}(\vec{c}_1 - \vec{c}_2)$. Across the core of the dislocation, the crystal structures are related by the translation $\vec{b}$.

In chapter 4 an example of a *stacking fault* has been described in an FCC structure. It is due, for instance, to the lack of an atomic plane in the sequence ABC (Fig. 7.9 (a)).

A crystal defect combining two parallel dislocations D_1 and D_2 *and a stacking fault* can be energetically more favourable than a single dislocation D. The occurence of such a situation is the *dissociation* of D into two dislocations. However, the dislocations D_1 and D_2 are *imperfect dislocations*, a type of linear defect not yet considered in the preceding chapters.

Assume that in a B-plane one has two parallel lines D_1 and D_2 along each of which a discontinuous displacement has taken place. Their interval is a ribbon of width W, in which the atomic configuration is that of a C-plane (Fig. 7.9 (c)). The translation which transforms a B-structure into a C-structure is equal to the vector $\vec{b}_1 = \vec{g}$ already recalled. Likewise, the translation which transforms the C-structure into a B-structure is equal to $\vec{b}_2 = \frac{1}{6}(2\vec{c}_1 - \vec{c}_2 - \vec{c}_3)$ (Fig. 7.9 (b)). The two lines D_1 and D_2 are similar to dislocations since they are lines of discontinuity separating two regions whose structures are related by a translation shift. However, $\vec{b}_1$ and $\vec{b}_2$ being distinct from primitive translations of the structure, D_1 and D_2 are necessarily associated to a defective portion of planar surface (cf. chapter 5, section 5.2.1). This surface is constituted by the ribbon of width W. D_1 and D_2 are called *partial dislocations*.

Note that if one overlooks the ribbon W, the pair of dislocations D_1, D_2 has the same effect within the structure of the solid, as a single dislocation D with Burgers vector $\vec{b} = (\vec{b}_1 + \vec{b}_2)$. Indeed, such a "thick" dislocation is a narrow region of discontinuity across which part of the crystal (the left part of the B-plane) has slipped by $\vec{b}$, with respect to the other part (the right region of B). The dislocation D can be considered to be *dissociated* into D_1 and D_2.

Such a dissociation can be energetically advantageous. Indeed the following inequality holds:

$$\vec{b}^2 = \frac{c^2}{2} > \vec{b}_1^2 + \vec{b}_2^2 = \frac{c^2}{3} \tag{7.21}$$

Since the elastic energy induced by a dislocation in the solid is proportional to the square of the modulus of the Burgers vector, it appears that the pair of partial dislocations (D_1, D_2) is associated to a smaller energy than D. However, the presence of the stacking fault also has a positive contribution to the energy of the solid. Clearly, this contribution will increase with the width W. Moreover, there is an interaction between the two partial dislocations, which as shown in chapter 6 is proportional to the *positive* quantity $\vec{b}_1 \cdot \vec{b}_2$. This repulsion will decrease as measure as the distance W between D_1 and D_2 increases.

Since the two latter contributions vary in an opposite sense as a function of W, their sum will determine an optimal value for the width W_0 of the ribbon separating the two partial dislocations (cf. exercise 19). For this value of the width, the total positive contribution can be smaller than the lowering of energy determined by Eq. (7.21). The dissociation of D into D_1 and D_2 will then occur.

Numerical simulations (Fig. 7.10) performed for metals having the fcc structure (e.g. aluminum), or the CC structure (e.g. iron), as well as available observations at the microscopic scale, have shown that in these systems, the dislocations are actually dissociated. For instance, in aluminum, the width of the stacking fault is found to be of the order of $\approx 5b$, i.e.

$\approx 20\text{Å}$. Such an order of magnitude corresponds to the size of the core of a dislocation. We can therefore consider that a dissociated dislocation is merely a dislocation whose core has a more complex structure than that considered in the above model.

This complexity does not modify the conclusions relative to the mobility of a dislocation in its glide plane. Indeed a dissociated dislocation corresponds to a wide core, an element which favours the easy displacement of the dislocation.

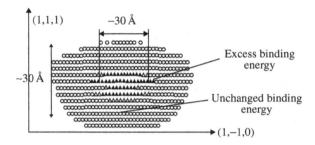

Figure 7.10 Result of the study, by numerical simulation, of the core of a dislocation in aluminum (A. Aslanides, thesis 1998). The white circles represent the atoms whose binding energy is the same as in the ideal crystal structure. The other symbols represent atoms whose binding energy is increased by an abnormal configuration of the bonds. The figure shows that the most perturbed region (black triangles) is dissociated, the partial dislocations being distant by 20 Angströms.

Chapter 8

Microscopic mechanism of plasticity

Main concepts: Relationship between plastic deformation and localized shear strains. Motion of dislocations. Dislocation sources. Physical origins of the yield stress and of the hardening. Plastic deformation rate. Influence of temperature. Origin of the diversity of plastic properties of materials.

8.1 Introduction

The results worked out in the preceding chapters are completed here, and then used in the view of understanding the general principles underlying the plastic behaviour of solid materials. The causes governing the differences between the plastic properties of different materials will also be addressed (e.g. the great ductility of copper versus the brittleness of glasses). The latter aspect is more complex to account for. In solids, this difficulty of going beyond a general understanding of a property is not specific of plasticity. Hence, it is easy to explain why metals are good conductors of the electric current, while other materials are electrical insulators. By contrast, it is more complicated to explain why copper or silver are better conductors than other metals, and furthermore, to calculate their respective conductivity. This general difficulty comes from the fact that the physical properties of a solid are the result of interactions between a very large number of constituants. In the case of plasticity, an additional difficulty is that two, very different length scales, are involved in determining this property: the interatomic distances (cf. chapter 7), and the characteristic length of the interactions between dislocations (cf. chapter 6).

Figure 8.1 reproduces the curve already shown in chapter 1, representing the standard mechanical behaviour of a solid. The main aspects of the *plastic* range of this curve will now be considered from a microscopic standpoint. In this view, several topics must, successively, be clarified.

First, the crystallographically *discontinuous* character of the plastic deformation is emphasized. The nature of the yield strength (or elastic limit) σ_Y which characterizes the *hardness* of a solid is then analyzed. It appears that the relatively low values of the hardness of many metals is due to the presence, and ease of motion, of dislocations in these materials. The number of dislocations required to account for the magnitude of the currently observed plastic deformations can be estimated. This number is very large, and its magnitude implies that most dislocations are not present in the beginning of the deformation process, but that they are continuously produced *during the deformation.* Mechanisms of such a production will be described. The origin of the *cold work hardening,* i.e. the increase of the yield strength σ_Y induced by the plastic deformation itself, will then be explained (Fig. 8.1). The elements into play in this phenomenon are the same as the ones which determine the *ductility* of the material (measured by the magnitude of the interval YF in Fig. 8.1). Finally, the influence of temperature will be considered. Generally, an increase of temperature makes the plastic deformation easier. This effect is specially large in hard, and brittle, solid materials.

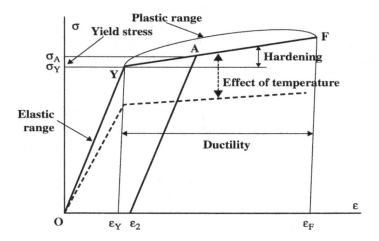

Figure 8.1   Reminder of the macroscopic aspects of the mechanical properties of a solid.

8.2 Plastic deformation and local shear

As indicated in the introductory chapter (cf. section 1.2.1),) the permanent elongation of a test-rod, submitted to a traction on its end-surfaces, consists of a set of *slips*, with "slices" of the rod, gliding upon the adjacent slices, parallel to a given plane direction. This direction is inclined, by an angle θ, with respect to the direction of the exerted traction. Thus, the deformation of the rod is not uniformly distributed in its volume. It is *localized* on the various glide planes. On either sides of such planes, the solid undergoes a *shear deformation*, in which a slice of the crystal is uniformly translated with respect to the adjacent slice. Figure 8.2 describes the details of this set of slips.

8.2.1 *Amplitude of the slips*

Each slip has an amplitude g comprised between 10Å and 1000Å. The spacing h between the glide-planes has values of an order comprised between a hundred Angströms and a micron. Such values account satifactorily for the observed elongations actually observed in the plastic deformation of materials, even in the ones having the largest ductility. Consider, for instance, a 10 cm long rod, in which the slips occur along planes inclined by 45° with respect to the axis of the rod. If their amplitudes are $g = 10^3$Å, and their spacing $h = 10^3$Å. the cumulated magnitude of the slips is ~7 cm. Taking into account the angular effect, it determines a relative elongation of the rod by ~50%.[1] In the most ductile materials, the observed elongation, which can reach relative values of the order of 1000%, would correspond to values of g and of h of the order of 100 Å.

8.2.2 *Resolved shear stress*

Although the external stress applied to the rod is due to an axial traction (e.g. σ_{33}), it is a shear stress σ_{cis}, denoted *resolved shear stress*, which induces in a direct manner the gliding of a portion of the solid on the adjacent portion. In order to express the relationship between σ_{33} and σ_{cis} as a function of the orientation of the glide plane (Fig. 8.2), let θ and ϕ' (with $\phi = \pi/2 - \phi'$) be respectively the angles of the $\vec{z}$-direction with

---

[1]As discussed in the next section, as the elongation develops, the direction of the glide planes becomes closer to the axis of the rod. The value of the effective elongation of the rod is then closer to the value of the cumulated glides.

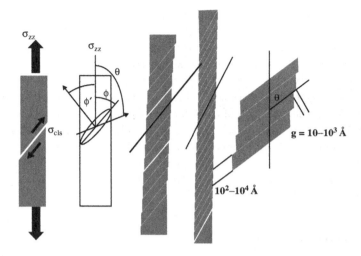

Figure 8.2 (Left to Right) 1°) beginning of the elongation of a rod submitted to an axial traction with the onset of a single slip of a part of the rod on the adjacent part. This effect is obtained by the application of an effective "resolved" shear-stress σ_{cis} resulting from the axial stress σ_{33}; 2°) Geometrical scheme corresponding to the relationship between the two stresses. 3°) Gradual evolution of the elongation of the rod by multiplication of the slips along parallel planes, and by increase of their amplitudes.

the *direction* $\overrightarrow{\gamma}$ *of the glide* and with the *normal* $\overrightarrow{n}$ *to the glide plane* . Equation (3.2) in chapter 3 determines the force exerted on the unit area of the glide plane:

$$\overrightarrow{f} = \overline{\overline{\sigma}}.\overrightarrow{n} = \sigma_{33}\cos\phi'.\overrightarrow{z} \qquad (8.1)$$

Its projection on the glide direction is $f_{cis} = f\cos\theta$. On the other hand, f_{cis} can be written as:

$$f_{cis} = \overline{\overline{\sigma}}_{cis}.\overrightarrow{n} = \sigma_{cis}. \qquad (8.2)$$

from which the value of the resolved shear stress can be deduced (Schmid formula)[2]:

$$\sigma_{cis} = \sigma_{33}.\cos\theta.\cos\phi' = \sigma_{33}.\cos\theta.\sin\phi \qquad (8.3)$$

As shown on Fig. 8.2 the application of an axial stress σ_{33} has the effect of inducing a progressive rotation of the glide planes towards the axis of the rod, in such a manner as to decrease the value of θ, and thus to increase

---

[2]The angles considered in some books are sometimes θ and ϕ' the resulting expression is then $\sigma_{cis} = \sigma_3 \cos\theta \sin\phi'$.

the value of $\cos\theta$. Hence, the elongation of the rod becomes progressively closer to the cumulated magnitude of the slips.[3]

8.2.3 *Crystallographic characteristics of the slips*

The experimental study (e.g. through X-ray or electron diffraction) of plastically deformed crystals shows that the regularity of the atomic configuration is, up to a very high accuracy, identical to that of the undeformed crystal. The implication of such an observation is that the slips do not induce *dammage* in the material, i.e. no *disorder* in its atomic structure. This is consistent (Fig. 8.3), on the one hand, with the fact that the slips occur along lattice planes, and, on the other hand, because the amplitudes g of the slips are exact multiples of the lattice translations in the direction of the slips. The slips modify slightly the external surface of the sample by generating on it small steps (cf. Fig. 8.3). However, in the major part of the volume of the sample, the ideal structure of the crystalline solid is preserved. This preservation also explains the fact that, after a plastic deformation, the slope of the "elastic" stress-strain curve $A\epsilon_2$ (Fig. 8.1), is the same as the slope of the initial elastic range OY, prior to the plastic deformation. Indeed, this slope is equal to an elastic stiffness coefficient which is only dependent of the atomic structure *in the volume* of the solid material. The elementary stage of a slip is the gliding of an atomic plane on the adjacent plane by a translation equal to *a single lattice translation* b. Such a slip determines for the parallelipipedic sample with edges A, B, C (Fig. 8.5 (c)) a shear of value:[4]

$$\gamma \approx \frac{b}{C} \tag{8.4}$$

In the deformation shear mode represented in Fig. 8.2, the slips occur along parallel planes belonging to a single family of lattice planes. Their spacings h are therefore also multiples of the spacing between adjacent planes.[5]

More generally, depending on the crystal symmetry of the solid, the slips can occur along one or several equivalent directions of lattice planes. These different directions correspond to symmetry-related lattice planes and, thus,

---

[3]Effects of this rotation of the glide planes in metals having the FCC structure will be discussed in section 8.9.2.

[4]As emphasized before, the shear strain component ϵ is equal to the half of the shear γ.

[5]Such a spacing is the projection, on the normal to the planes, of lattice translations of the crystal.

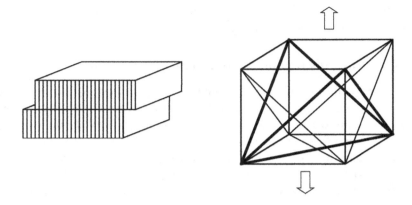

Figure 8.3   (Left) Shear corresponding to the global gliding of half of the crystal. Its magnitude is equal to a multiple of a lattice translation. (Right) In a FCC crystal, directions of glide planes which can be activated simultaneously by an axial traction of the rod.

to an identical spacing between adjacent planes. An example is provided by the various atomic planes of an FCC cubic structure, perpendicular to the diagonals of the cubic cell (Fig. 8.3). If the axial traction of the sample induces a resolved shear stress of sufficient magnitude on several of these plane directions, slips will occur *simultaneously* along these directions. Figure 8.4 shows an example of the effects of slips in a single direction of planes and in several directions.

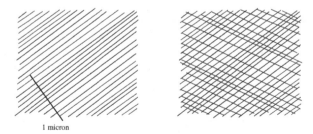

1 micron

Figure 8.4   Schematic reproduction of an experimental observation (mentioned in the reference book by Adda *et al* quoted in chapter 1) of the surface of an aluminum test-sample submitted to an axial stress. The lines correspond to surface steps induced by slips parallel to lattice planes. The right part of the figure shows the result of the activation of two equivalent glide systems.

8.3 Elementary slip and yield strength

The above description shows that the plastic behaviour is partly governed by the mechanism of slips.

The yield strength of a solid is the minimum stress required to induce a non-reversible deformation. As just stated, this stress is the one required to induce the slip of a portion of the solid on the adjacent portion, parallel to a lattice plane.

(a)   (b)   (c)

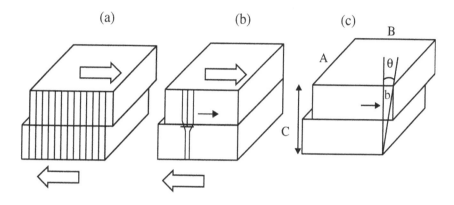

Figure 8.5   (a) Result of an ideal (global) shear of amplitude b, (b) Partial slip, limited to the edge-dislocation line. The displacement of the dislocation to the right of the sample leads to its evacuation, with creation (c) of a step of width b on the surface and is thus equivalent to the result obtained in (a).

Two mechanisms producing a slip parallel to a lattice plane can be considered. One is the uniform translation of the entire "upper" part of a crystal (Fig. 8.5 (a)) on the adjacent part. One has then to deal with an *ideal shear*. The other is the partial and progressive slip induced by the motion of a dislocation in its glide plane across the sample.

8.3.1 *Edge dislocations*

Figure 8.5 (b) shows the displacement of an edge dislocation in its glide plane, parallel to its Burgers vector. Remind that the dislocation is the discontinuous line which separates the upper portion of the solid situated at its left, displaced by the translation $\vec{b}$, and the solid situated at its right which has not been displaced. When the dislocation moves to the right a larger fraction of the upper half of the crystal undergoes the translation $\overrightarrow{b}$. When the dislocation reaches eventually the surface limiting the crystal at its right, and is thus evacuated, the entire upper half of the crystal has been translated by $\overrightarrow{b}$, and a step of amplitude b is formed on the surface. The gradual motion of the dislocation across the width of the sample has therefore the same final effect as a uniform translation of the entire upper half of the crystal, i.e. as an ideal shear.

If n such dislocations move across the crystal in the same glide plane, the resulting step on the external surface will be nb. Figure 8.6 shows the result of the motion across the sample of several dislocations gliding in parallel planes: a series of steps having various widths arise on the surface of the sample.

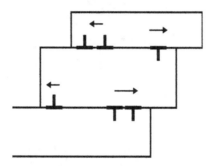

Figure 8.6   Steps on the surface of a sample, resulting from the motion and evacuation of dislocations situated in parallel glide planes. A given step can be due to a "positive" dislocation or to a "negative" one moving in the opposite direction. The two motions are obtained for the same orientation and sign.

8.3.2   *Screw dislocations*

Figure 8.7 decomposes the similar elementary process in the case of a screw dislocation. Again, the evacuation of the dislocation produces a step of amplitude $\overrightarrow{b}$, but situated on the external surface AB.

It is worth pointing out that the occurence of such steps is related to a specific *growth mode* of crystals, i.e. a process by which a solid is formed out of the liquid phase (for instance, during the freezing of the liquid). Thus, consider a crystal formed by the gradual aggregation of atoms present in the liquid (melted solid or solution). If a step, induced by a screw dislocation, exists on the surface of the small crystal sample at an early stage of its growth, additional atoms will aggregate to the sample preferentially on the step. Indeed, on a step, an atom is bonded to the solid by *two bonds* (on the surface and on the step), while at other sites of the crystal, a single bond exists. The former configuration of bonds, has less energy since it

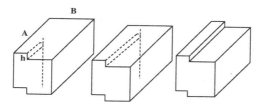

Figure 8.7   Step induced on the surface perpendicular to a screw-dislocation line during the gliding-motion of the dislocation. A similar step is also induced on the opposite face of the sample.

is closer to the equilibrium configuration of the ideal crystal (in which an atom is entirely surrounded by neighbouring atoms). Hence, the growth of the crystal will proceed by a widening of the step. Since the step is bound to the screw-dislocation line, this widening will have the effect of forming an inclined spiral of height b. Figure 8.8 shows the progressive formation of this spiral, whose observation on the surface of the sample is an evidence of the existence, in the crystal, of a screw dislocation.

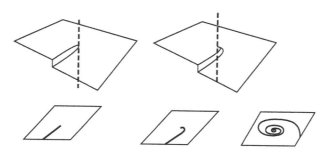

Figure 8.8   Formation of a growth spiral around the point where a line of screw dislocation emerges from the surface of the crystal.

8.3.3   *Upper limit of the yield strength*

The upper limit of the yield strength (also termed the *ideal shear strength*), determined by the bonding between atoms, has been evaluated in chapter 7 (Eq. (7.10)). It is equal to the shear stress required to create a first dislocation. Its value is:

$$\sigma_{max} = \frac{\mu b}{2\pi a} \approx \frac{\mu}{10} \approx 10^6 \text{Pascals} \approx 1 \text{ Ton/mm}^2 \qquad (8.5)$$

in which μ is the shear modulus, a the lattice translation normal to the glide plane, and b the lattice translation parallel to the glide direction. Such a stress corresponds to a strain value of the order of 10%. Yield strength values having this order of magnitude have been measured in so-called "whiskers", i.e. very thin crystals, of diameter of the order of a few microns, which have an almost perfect crystal structure and do not contain any dislocation.[6]

However, the situation generally observed in solids corresponds to yield strength values less than $10^{-3}\mu$ and which can even be as low as $10^{-6}\mu$ in certain metals such as copper or aluminum. Such values are not compatible with the upper limit (8.5). As shown in the next paragraph, they are in agreement, instead, with a deformation mechanism "assisted" by the motion of dislocations moving across the sample, in a direction parallel to their glide plane ($\overrightarrow{b}$, $\overrightarrow{d}$).

8.3.4   *Slip assisted by dislocation motion*

The stress needed to produce the displacement of an isolated dislocation in a crystal, parallel to the Burgers vector, is the Peierls-Nabarro stress defined in chapter 7 (cf. section 7.3.2), $\sigma_{PN} = 2\mu \exp(-\gamma w/b)$ with $\gamma \tilde{} \pi$. The resulting value of σ_{PN} for different values of the core-width of a dislocation, is summarized in Table 8.1.

Table 8.1

$w \rightarrow$	b	$2b$	$3b$	$4b$
$\sigma_{PN} \rightarrow$	$\frac{\mu}{12}$	$4.10^{-3}\mu$	$10^{-4}\mu$	$7.10^{-6}\mu$

$$\qquad (8.6)$$

---

[6]The reason for the absence of dislocations is due to their small diameter. If dislocations were present they would be attracted by the cylindrical surfaces, and are therfore evacuated from the sample. The attraction by the surfaces has its origin in the fact that the energy of a dislocation, being $\propto \mu b^2$, has a tendency to migrate towards a region in which μ has a small value.This is the case of the region close to the surface where $\mu_{surf} \ll \mu_{vol}$.

The yield strength in each case will be equal to the corresponding σ_{PN} value. It appears that the values in Table 8.1 actually agree with the experimental ones measured in real materials.

As shown in section 7.2.4, the core of a dislocation is generally wide if the glide plane is a dense atomic plane of the structure. It is therefore the value of σ_{PN} relative to the denser atomic planes of a structure (e.g. the smallest σ_{PN}) which will determine the yield strength of the material.

Note that σ_{Pn} is the stress required to put dislocations into motion, and the former argument thus holds only if dislocations are *already present* in the solid. If, however, a first dislocation must be created, the yield strength will have to induce the slip of part of the crystal on the other part, and is therefore equal to the upper limit of the yield strength (Eq. (8.5)), of the order of $\mu/10$.

The presence of dislocations in a "virgin" sample is believed to have its origin in the very strong internal stresses (i.e. stresses exerted by one part of the solid on the adjacent parts) which develop during the process of solidification of the sample, from its melted state, and its cooling. Large temperature gradients between the surface and the inner volume are then present. The resulting inhomogeneous thermal expansion of the different regions of the volume generate internal stresses which can be sufficient to introduce dislocations.

8.3.5 *Number of dislocations*

In order to obtain a 100% elongation of a sample of centimetric size, with $b \approx 2,5\text{Å}$, and an inclination of the glide planes of 45° with respect to the direction of traction, it is necessary (cf. Eq. 8.4 with $C = 1$cm) that approximately 10^9 dislocations be displaced across the sample. In a metal which has not yet been deformed, the number of dislocations currently observed is in the range 10^4cm^{-2}-10^7.cm^{-2}. Moreover, for reasons which will be discussed in sections 8.4.2 and 8.4.3, only a fraction of these dislocations can be put into motion by an applied stress. Hence, the number of preexisting dislocations required to explain the observed ductility in many metals is not large enough, by several orders of magnitude. Note also that the dislocations being evacuated during the plastic deformation, their number should decrease in the course of the deformation process. Actually, the opposite effect is observed. As shown in Fig. 8.9, a strong increase of the number of dislocations is generally observed after repeated plastic deformations.

The former argument suggests that, in addition to the dislocations already present in a solid, whose existence satisfactorily explains the low value of the yield strength, a large number of dislocations are produced during the course of the plastic deformation.

Figure 8.9   Electron microscopic image of a sample of alloy TiAl after a plastic deformation by 2%. The number of dislocations has strongly increased (from $10^6/cm^2$ to $10^{12}/cm^2$), and a number of junctions are formed (Photograph from the thesis of F.Ferrer, 2000).

8.4   Dislocation sources

Several possible mechanisms for *generating continuously new dislocations*, during the plastic deformation of a solid, have been considered theoretically, and their relevance confirmed experimentally.They all involve the application of a small stress to pre-existing dislocations. The main one, the so-called *Frank-Read mechanism*, is analyzed in detail in the next section. Another one, the "spiral-source" is discussed more briefly. The Frank-Read mechanism relies on two properties. The first one has already been invoked in section 6.4: A shear stress acting on a segment of a dislocation whose two end-points are pinned (cannot be displaced), induces a curvature which increases with the value of the stress. The second property is the fact that an attractive interaction between two parallel portions of a single dislocation can lead to a "reaction" in which the dislocation is decomposed into two distinct dislocations.

8.4.1   *Change of configuration of dislocations*

It was shown in chapter 6 (section 6.5) that two parallel straight (edge- or screw-) dislocations with opposite Burgers vectors, attract each other and can annihilate if one is free to move towards the other. Consider two dislocations of arbitrary shape (Fig. 8.10-(left)), having opposite Burgers vectors $\overrightarrow{b}$ and $-\overrightarrow{b}$, and which possess quasi-parallel segments.

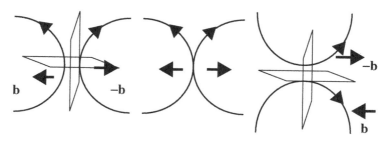

Figure 8.10   The central portions of two dislocations having opposite Burgers vectors attract each other and eventually annihilate, thus forming two other dislocations. The dotted rectangles are Burgers circuits.

As in the case of straight dislocations (section 6.5), the segments will attract eachother, and can mutually annihilate.

For instance, if these segments are edge-type, they are associated to two "half stripes" of additional lattice planes. The two segments move towards each other (Fig. 8.10-(center)) and eventually annihilate. In this annihilation, the two half-plane stripes join, and determine a single continuous stripe. This effect must not modify the closure defect of the various Burgers circuits drawn in the solid. As a consequence, the remaining sections of the two dislocations will join in order to form two new dislocation lines having respectively Burgers vectors $\overrightarrow{b}$ and $-\overrightarrow{b}$, and other shapes (Fig. 8.10 (right)).

8.4.2   Frank-Read source

Consider (Figs. 8.11 (a), 11(b)) a segment AB of a straight-edge-dislocation D, oriented by the vector $\overrightarrow{d}$, and of Burgers vector $\overrightarrow{b}$, ending on two dislocations D_1 and D_2 whose directions are situated out of the glide plane of D. Anticipating on the discussion of section 8.5, assume that D_1 and D_2 are locked at their position in the solid and cannot move. The segment AB, in attractive interaction with D_1 and D_2, is then "pinned" at the two fixed-nodes A and B. The remaining length of AB is assumed to be mobile in its glide plane $(AB, \overrightarrow{b})$.

A shear stress σ parallel to the glide plane (Fig. 8.11 (a)) is applied. In agreement with the result of section 6.4, the force acting on AB, perpendicular to the dislocation line on each of its points, induces as shown by Fig. 8.11 (c) a curvature of radius $R = T/\sigma$, in which $T = \alpha\mu b^2$ is the line tension of the dislocation line. Since the nodes A and B are fixed, the smallest *equilibrium* radius (i.e. the maximum curvature) $R_0 = (AB/2)$ (Fig. 8.11 (d)) is reached for an applied stress equal to $\sigma_0 = (T/bR_0)$ (cf. section 6.4). If the applied stress is further increased beyond this value, the line tension of the dislocation is not sufficient to balance the external forces exerted on the different portions of the half-circle AB. An instability takes place affecting the shape of the dislocation. Since the external forces are always perpendicular to each portion of AB, they will give rise to the succession of shapes represented on Figs. 8.11(e) and 8.11(f) in which the end points A and B remain fixed.

In the configuration represented on Fig. 8.11 (f) two portions of the dislocation, ab and cd, facing eachother are almost parallel. Since they both belong to D, they possess the same Burgers vector. However, their directions $\overrightarrow{d}$ are opposite. Referring to chapter 5, section 5.2, such a situation is physically equivalent to having two parallel portions of dislocation *with the same director* $\overrightarrow{d}$, *but opposite Burgers vector*. These dislocation segments therefore attract eachother. In agreement with the argument given in the preceding paragraph, and illustrated in Fig. 8.10, an annihilation will occur leading to a decomposition of the dislocation. This "reaction" gives rise to two dislocations. On the one hand, a small segment AB whose end points are pinned, and, on the other hand, a closed dislocation loop B_1. Both possess the same Burgers vector $\overrightarrow{b}$ (Fig. 8.11 (g)).

The direction of the force (Eq. (6.22)) exerted by σ on the two resulting dislocations tends to enlarge the diameter of the loop B_1, which acquires a circle-like shape whose diameter increases as a function of time. It also brings the curve AB back to the straight shape it had initially.

Pursuing the application of the stress, the new segment AB will give rise, by repeating the above evolution, to a series of loops B_2, B_3, similar to B_1. As measure as the radii of the loops increase, certain sections will be evacuated through the surfaces of the solid. Eventually, a succession of equally spaced quasi-straight dislocations will remain, moving through the solid. For instance, if the lines which have not been evacuated are vertical lines, as shown in Fig. 8.11 (h), they will be of edge-type, with "positive" edge-dislocations moving to the right and "negative" ones moving to the left. Each such dislocation will produce, on being evacuated through the surface, an increment b of plastic deformation.

Hence, this process describes a *dislocation source* which generates an infinite number of dislocations moving throughout the solid.

As a quantitative illustration, if $AB \approx 1$ micron and $b \approx 2.5\text{Å}$, the minimum stress $\sigma_0 = (2\alpha\mu b/AB)$ which must be applied to start the generation of the series of dislocation loops is in the range $10^{-3} - 10^{-4}\mu$. These values are of the same order of magnitude as the yield strength (cf. section 8.5). One can therefore expect that this mechanism of generation will be activated as soon as the solid starts to deform plastically. Note that a large value of the line tension $T = \alpha\mu b$ will be an obstacle to the Frank-Read mechanism.

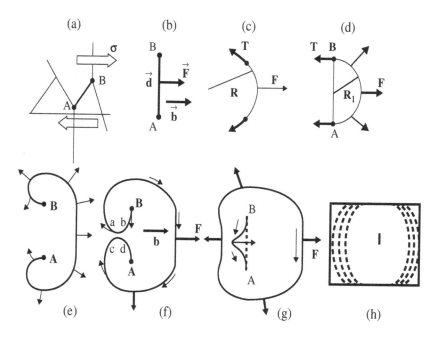

Figure 8.11 Generation of a sequence of dislocations by the Frank-Read mechanism. (a) and (b) represent respectively the dislocation in space and in projection on the glide plane $(\vec{b}, \vec{d})$. (h) is an overall view of the sample showing, at another scale, the two pinning points A and B, and the successive dislocations generated by the source and emitted in both senses towards the surfaces of the sample under application of a stress.

8.4.3   *Spiral source*

A so-called "spiral source" is represented in Fig. 8.12. It is initiated by a segment of a straight edge-dislocation line comprised between a fixed node A (again, pinned on the intersection with a non-mobile dislocation) and a surface of the sample.

A shear stress parallel to the glide plane will, as in the Frank-Read source, induce a force perpendicular to the dislocation which will have two effects. First, it will tend to rotate the segment around the fixed node A. Second, it will induce progressively the shape of a spiral. Indeed, the *linear velocity* of the portions of the dislocation is likely to be constant along the length of the line. Hence, the *angular velocity* will be largest near the node A, and deform the line into a spiral shape. Simultaneously, the various open loops of the spiral move away from A.

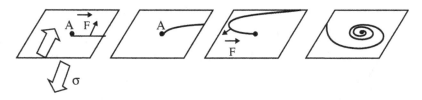

Figure 8.12   Spiral source of dislocations. The branches of the spiral rotate and move away from the node A under the force exerted by the applied stress. When they reach the surface of the sample, they are evacuated and create plastic deformation increments of amplitude b.

Whenever the most external portion of the spiral "sweeps" the surface of the sample, it is evacuated, and an increment b of plastic deformation is produced. The continuous rotation of the spiral and its enlargement, under the action of the applied stress, thus generates a plastic deformation whose value increases as a function of time.

8.5   Dislocation glide and climb

The mechanisms considered in the preceding paragraphs rely on the mobility of a dislocation *in its glide plane*. It is possible (cf. chapter 7) to set into motion an edge dislocation in its glide plane $(\overrightarrow{d}, \overrightarrow{b})$, parallel to its Burgers vector, by applying a stress sufficient to overcome the energy barrier existing between two Peierls valleys (i.e. stable symmetric locations of the dislocation line separated by a distance b). The motion across the entire sample will then be achieved by a succession of elementary jumps over a distance b. The same mechanism holds for a screw dislocation. In this case, any plane containing the dislocation is a glide plane ($\overrightarrow{b}$ and $\overrightarrow{d}$ being parallel). The gliding of a dislocation *does not involve any supply or evacuation of solid matter* (i.e. atoms) from the surrounding crystal.

There is another simple mechanism of displacement of an edge dislocation, the *climb*, which, by contrast to the glide, requires a displacement of matter (atoms) from one region of the solid to another. At the atomic level, this "transport" of matter is achieved through *the diffusion of atoms or of vacancies* (cf. chapter 4).

Figure 8.13 shows the migration of a vacancy into the atomic row limiting the additional half-plane of an edge dislocation. Equivalently, an atom of this row has jumped into the neighbouring empty site of the vacancy. Schematically, the vacancy has the effect of creating two "jogs" (i.e. two segments of dislocation, perpendicular to the straight line, which are the first step of an upwards vertical shift of the dislocation line, normal to the glide plane. If many vacancies migrate and substitute all the atoms of the considered row, the dislocation will have globally been displaced upwards by one interatomic distance.

Conversely, if the migration concerns interstitial atoms which migrate to occupy atomic sites below the last row of atoms limiting the additional half-plane, the dislocation will be displaced downwards.

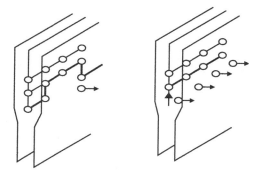

Figure 8.13   Climb of an edge-dislocation assisted by the migration of vacancies towards the lower limit of the additional half plane.

These two types of displacements, which occur perpendicularly to the glide plane, consitute a *climb* of the dislocation.

The driving force for this mechanism is the attraction of vacancies by the "compressed" region neighbouring the edge-dislocation (cf. section 6.6), or the attraction of interstitial atoms by the "expanded" region neighbouring the dislocation. The climb requires an efficient migration through the mechanism of diffusion in the solid. As shown in chapter 4, the diffusion coefficient of vacancies or of interstitials only becomes significant at high enough temperatures, e.g. values of the order of the melting temperature. Hence, *at "low temperatures", only the motion in the glide plane is possible.* At high temperatures, both types of motions (glide and climb), will be active to determine the plastic behaviour of a solid.

8.6 Hardening

The hardening of a material consists in the fact that, subsequent to one or several plastic deformations, the yield strength increases (point A of the curve $\sigma(\epsilon)$ in Fig. 8.1). This effect has its origin in the *increase of the density of non-mobile dislocation.* Their presence is an obstacle to the motion of mobile dislocations. Indeed, as shown hereabove, the process of plastic deformation implies the generation (through the activation of dislocation sources) of new dislocations. A fraction of these dislocations will contribute, by being evacuated from the sample, to increments of the plastic deformation. Another fraction will be pinned to the crystal, and therefore become non-mobile. Several mechanisms can induce this pinning.

8.6.1 *Pinning dislocations*

Various types of interactions between dislocations, or between a dislocation and another type of defect, can prevent a dislocation from moving.

Interaction between two dislocations

In chapter 6, the only interaction considered in detail concerned two parallel dislocations sharing the same glide plane. It was mentioned, however, that in general two dislocations of arbitrary type and direction are likely to exert on one another a set of forces. If one of these dislocations is, for some reason, pinned to the lattice, its interaction, whether attractive or repulsive with a mobile dislocation, will have the effect to oppose its motion across the sample. The dislocation will become *sessile*, i.e., parts of it will be locked at their position in the crystal. Figure 8.14 illustrates this situation which constitutes the main cause opposing the displacement of mobile dislocations.

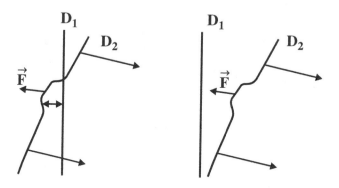

Figure 8.14 Obstacle to the motion of a dislocation due to its interaction with a non-mobile dislocation. (Left) Case of a repulsive interaction. (Right) Attractive interaction. In both cases the force exerted is opposed to the motion.

Pinning by a junction

Whenever two dislocation lines, having Burgers vectors $\vec{b}_1$ and $\vec{b}_2$, cross, their interaction can produce, if $\vec{b}_1 . \vec{b}_2 < 0$, a *junction*, which is an additional segment of dislocation (Fig. 8.15 (b)). Equation (5.2) shows that the junction will have the Burgers vector $\vec{b} = \vec{b}_1 + \vec{b}_2$ provided that the different dislocations are oriented in continuity.

If one assumes that the total length of the two dislocation lines is preserved, the energy of the junction will be proportional to $b^2 = (b_1^2 + b_2^2 + 2\vec{b}_1 . \vec{b}_2)$. Its value will be less than than the energy of the portions of the initial dislocations having the same length.[7]

Let AB be a junction (Fig. 8.16 (a)) whose glide plane is P, and whose end points are nodes situated on two dislocations having glide planes distinct from P. Although none of the segments involved are pinned to the lattice, their interaction can prevent them from moving for a variety of reasons:

---

[7]This is the same type of "reaction" described in section 8.4 for $\vec{b}_1 = -\vec{b}_2$. In the latter situation, the hypothetical junction would have the Burgers vector $\vec{b} = 0$, which corresponds to the absence of a dislocation.

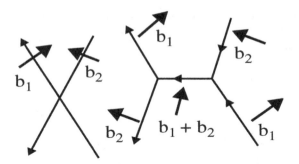

Figure 8.15   Formation of a junction whose energy is lower than that of the initial configuration.

i) If the glide directions of the three dislocations are not compatible.Thus, the stress applied has an orientation such as to put into motion D_1 or D_2, but is not suitably oriented to displace AB. As a consequence, the interaction which has determined the existence of the nodes, opposes the displacement of D_1 and D_2, through the absence of motion of AB.

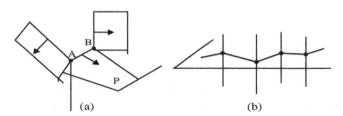

Figure 8.16   (a) Junction of several dislocations possessing distinct glide planes. (b) "Forest" of vertical dislocations intersecting different straight segments of a dislocation situated in the horizontal plane.

ii) The magnitude of the applied shear stress is sufficient to displace D_1, but is too small to displace AB, because the Peierls-Nabarro stress is larger for the dislocation AB than for D_1.

In any of these cases, the overall translation of D_1 or of D_2 will be prevented.

Figure 8.17 shows an example of junction observed by electron microscopy.

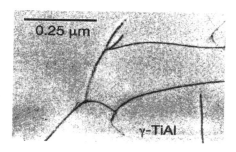

Figure 8.17   Observation, by electron microscopy of a junction in the metallic alloy TiAl. (Picture kindly provided by Yu-Lung Chiu, University of Auckland).

Pinning by the dislocation "forest"

A generalization of the preceding situation occurs when a dislocation D is constituted by a succession of straight segments, located in the same glide plane, whose various nodes are intersections with a series of other dislocations, almost perpendicular to the plane, and constituting a so-called "forest" of dislocations (Fig. 8.16 (right)).

Such a forest arises in FCC structures. Its existence is associated to an "equilibrium lattice" of dislocations, formed by straight dislocation segments, and called the *Frank lattice*.

Indeed, consider the most dense atomic planes of this structure. They are perpendicular to the main diagonals of the cubic cell, as, for instance, the plane perpendicular to the (111) direction.   In this plane, the shortest

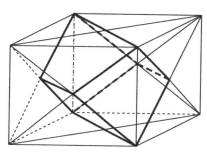

Figure 8.18   Frank lattice formed by segments of dislocations located at the intersection of dense planes of an FCC structure, and generating a three-dimensional lattice of dislocations.

Bravais lattice translations are the six primitive translations (cf. Eq. (2.9)), $\pm(1/2)(\overrightarrow{c}_i - \overrightarrow{c}_j)$ with $i \neq j = 1, 2, 3$. These are possible Burgers vector values, and it is possible to select three vectors, $\overrightarrow{b}_1$, $\overrightarrow{b}_2$, $\overrightarrow{b}_3$, in the former set having a sum equal to zero. In agreement with Eq. (5.2), the three vectors can be associated to three dislocations-lines located in the dense planes, meeting at a node. Due to the symmetry of the structure, these dislocation lines will make an angle of $2\pi/3$ with one-another. Hence, one can have, in each dense plane, a hexagonal lattice formed by dislocation segments (Fig. 8.18). The onset, in each dense plane, of such a lattice, determines a three-dimensional lattice of straight segments of dislocations, the *Frank lattice*, whose obervation has been asserted in certain materials (Fig. 8.19).

A dislocation moving in a dense plane will meet the forest constituted by the segments of the Frank lattice which are almost perpendicular to the plane.

Figure 8.19   Picture of a Frank lattice (reproduced from the web site of the University of Kiel, Germany).

Pinning by a jog

A jog is a short junction (its length is of the order of magnitude as that of the Burgers vector b), which is generated whenever two dislocations cross. As shown in Fig. 8.20, the translational shift of $\overrightarrow{b}$ which exists between the two parts of the crystal, on either sides of one of the dislocations, has the effect of decomposing the glide plane of the other dislocation. It determines

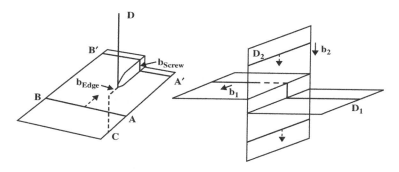

Figure 8.20 (Left) Formation of a jog of height b_{Screw} on an edge-dislocation AB, by crossing AB with a screw dislocation CD. A jog equal to $\overrightarrow{b}_{Edge}$ is also formed on the screw dislocation. (Right) Formation of a jog on an edge dislocation by the crossing of another edge dislocation.

two parallel half planes joined by a portion of plane perpendicular to the half planes, and containing the jog. The figure illustrates the formation of a jog when either an edge and a screw dislocation cross, or when two edge dislocations cross.

The jog, being a section of the initial dislocation, has *the same Burgers vector*. Indeed, the discontinuity of the displacement induced by the dislocation will not be changed. For this reason, the jog formed on an edge dislocation line, will remain of the edge type. By contrast, the jog formed on the screw dislocation will be of the edge type since its Burgers vector is perpendicular to its direction.

The creation of a jog can pin the motion of a dislocation. Consider the case of a screw dislocation D_s containing a jog, formed by crossing an edge dislocation (Fig. 8.21 (a)). As already emphasized, this jog is of the edge type. It can only glide parallel to the direction of D_s. Thus, at "low temperatures" it cannot move in a direction such as $\overrightarrow{u}$, perpendicular to $\overrightarrow{b}$ (such a motion would be a climb, only possible at high temperatures). The jog will therefore prevent the motion of the entire screw dislocation to which it is tied. The jog is a *sessile* (non-mobile) element of the dislocation.

Pinning by a grain boundary

Other types of microscopic objects are likely to prevent the motion of a dislocation, for instance a precipitate, or a grain boundary (cf. section 4.1). Their effect is to perturb the regularity of the crystal lattice and hence *to interrupt the continuity of the glide plane.* Shearing part of the crystal, which is required to propagate the dislocation, is made more difficult.

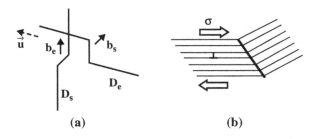

(a)   **(b)**

Figure 8.21   (a) Jogs induced during the crossing of a screw dislocation D_s and an edge dislocation D_e. The jog on the screw dislocation can only move by a climb mechanism in the $\vec{u}$ direction. It is termed "sessile". By contrast, the jog on the edge dislocation can glide. (b) Grain boundary opposing the free motion of an edge dislocation.

Figure 8.21 (b) shows schematically the boundary between two crystal grains. In each grain, has been represented the family of lattice planes for which the Peierls-Nabarro stress is smallest. The fact that the structures of the two grains do not have the same orientation, has the consequence that a given stress σ, applied to the "polycrystalline" sample, does not generate the same *resolved shear* σ_{cis} on the glide planes of the two grains. It can be sufficiently large to put the dislocations into motion in one grain ($\sigma_{cis}^1 > \sigma_{PN}$), and not large enough ($\sigma_{cis}^2 > \sigma_{PN}$) in the adjacent grain. The Boundary will then constitute a region of pinning of the dislocations, since their gliding is not possible in one grain.[8]

8.6.2   *Hardening mechanism*

The stress required to induce the motion of dislocations through the entire sample, and thus determine increments of plastic deformation, is equal to the sum $\sigma = \sigma_{PN} + \sigma_{obs}$. The Peierls-Nabarro stress σ_{PN} is the "intrinsic" yield strength in the absence of various obstacles, mentioned in the preceding paragraphs. σ_{obs} is the stress necessary to overcome the action of the various obstacles opposing the motion of dislocations (non-mobile dislocations, grain boundaries, precipitates, etc...). As explained, the various

---

[8]This is only one, simplified, of the mechanisms of pinning by a grain boundary. The boundary can have a disordered atomic structure, and its width can be much larger than a single atomic layer. The interaction with a dislocation is then more complicated.

pinning mechanisms will produce a progressive increase of the number of non-mobile dislocations in the sample, and thus, through this cumulatice effect, an increase of the value of σ_{obs}. Such an evolution is perceived, at the macroscopic level, as a *hardening* of the material. As already stated, direct observations by electron microscopy, performed before and after an expriment of plastic deformation, confirm that there is an increase of the number of dislocations by several orders of magnitude (e.g. from $10^6/cm^2$ to $10^{12}/cm^2$), as well as of their entanglement (cf. Fig. 8.9).

In solid materials in which σ_{PN} has a small value (e.g. metals with the FCC structure), the yield strength will be determined by σ_{obs} even in a "virgin" sample which has a small density of defects. Besides, as this density is likely to vary widely as a function of the "thermal and mechanical history" of the sample, the value of the yield strength will also vary in a wide range of values, for a given solid composition and structure. For instance, this value will be smaller if the sample has been *annealed* (slowly cooled) from a high temperature in order to obtain a crystal structure close to the ideal one (cf. section 8.7). In such a material, the slope of the hardening curve (cf. Fig. 8.1) will be large, since $\sigma \tilde{~} \sigma_{obs}$ increases with the increase of the density of obstacles.

8.6.3 *Unpinning of the motion of dislocations*

Whenever the gliding of a dislocation is blocked by an obstacle (pinning by the lattice, interaction with other dislocations or grain boundaries etc...), setting it again into motion can be obtained by increasing the applied stress, or by raising the temperature of the material. These effects will be examined in section 8.7. Other "internal" mechanisms can also be active.

Cross slip

Consider, for instance, a crystal with the FCC structure. As shown in Figs. 8.3 and 8.4, several equivalent families of dense lattice planes, having different directions exist (e.g. planes perpendicular to the cube diagonals). A *screw* dislocation located in one of these planes can be parallel to the intersection between two equivalent planes A and B. If this dislocation is blocked by an obstacle in its initial glide plane A, it can change direction of motion and glide along B, if the resolved stress in this plane is sufficiently large to activate its motion. The displacement of the dislocation is then a *cross slip* (8.22).

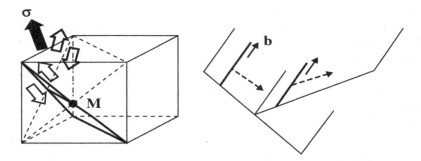

Figure 8.22 Cross slip for a screw dislocation parallel to the intersection of two dense planes of an FCC structure. The direction of the stress σ induces resolved shear stresses in the two dense planes. The dislocation is likely to move by gliding in either of the two plane directions.

Amplification of the external stress

Another mechanism consists in the process of "amplification" of the applied stress by the accumulation of dislocations succeeding eachother in the same glide plane (as for instance, the dislocations produced by a Frank-Read source), the "head" dislocation being blocked by an obstacle (see also exercise 31).

Figure 8.23 shows edge-dislocations piled-up behind a grain boundary. They all have the same sign, and therefore repell each other.

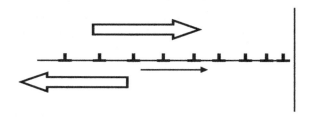

Figure 8.23 Set of dislocations produced by a source situated at the left of the figure. Its driving force is a shear stress represented by thick arrows. The dislocations are blocked in their motion by a grain boundary (vertical line) and pile-up in front of it.

Each dislocation is in equilibrium under the action, on the one hand, of the repulsive forces exerted on it by all the other dislocations, and, on the other hand, of the external stress which tends to displace it towards the obstacle. In addition the "head dislocation" is submitted to the blocking-force exerted by the obstacle. It is possible to analyze quantitatively this equilibrium using the equations expressing the interactions in section 6.6. This analysis is performed in exercise 31. Its results show that the stress field acting on the obstacle, in front of the head dislocation, and generated by the n dislocations and the external stress σ, is of the order of $n\sigma$. Hence the "piling up" of the dislocations in front of the obstacle determines an *amplification* of the applied stress. As a consequence, while σ is unsufficient to produce the motion of the head dislocation across the obstacle, the stress $n\sigma$ can be large enough to "shear" the obstacle and permit such a motion.

Figure 8.24 shows an actual example of a piling of dislocations behind an obstacle.

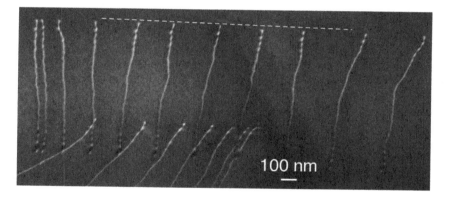

Figure 8.24   Series of dislocations in a nickel alloy. The dislocations at the left of the image are blocked by defects and prevent the motion from the dislocations arriving from the right. In the lower part of the image another "pile up" appears (kindly communicated by F. Pettinari-Sturmel, A. Coujou, N. Clément).

8.7   Effect of temperature

There are several effects of the temperature on the plastic behaviour of a material.

A first phenomenon is the activation of the "climb" of dislocations, due to an increased efficiency of the diffusion of vacancies. As already seen in chapter 4 (Table 2), the diffusion coefficient D_L, whose value determines the flow of vacancies is multiplied by 10^7 between 20°C and 1000°C. At high temperature, a vacancy can move by several microns per second. It will therefore be easier to remove or absorb a vacancy from a dislocation to produce an increment of "climb". Figure 8.25 (a) shows how the climb of an edge dislocation permits to avoid an obstacle consisting in a precipitate.

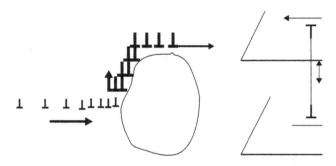

Figure 8.25   (a) Getting around a precipitate by a "climb" motion, activated at high temperature. (b) Two edge-dislocations of opposite sign, located in parallel glide -planes, can only anihilate mutually, if a climb motion is possible.

The activation of the climb can also lead to a reduction of the number of dislocations in the sample. Thus, if two edge dislocations having opposite Burgers vectors are located in parallel glide planes, their mutual attraction will displace them, as to face eachother at the smallest distance between the two planes. However it will not lead to their mutual annihilation unless a climb motion perpendicular to the planes can be activated (Fig. 8.25 (b)).

A second effect of a higher temperature is the reduction of the value of the Peierls-Nabarro stress σ_{PN}. Through thermal fluctuations, the temperature will increase the probability for the dislocations to cross the energy barriers between the Peierls valleys (section 7.3) and will therefore decrease the required value of the external stress. Note that the energy required being proportional to the length of the dislocation, the shortest the dislocation, the easiest the crossing of a barrier. For a "long" dislocation the crossing can proceed by steps, through creation of a *kink*, i.e. a short section of dislocation limited by two perpendicular jogs (Fig. 8.26). The multiplication of kinks and of jogs is not favourable energetically, because the length of the dislocation increases with the number of jogs. However, as measure as the temperature increases, the "equilibrium" number of short kinks increases, similarly to the increase of the number of point-defects in a crystal. Besides, a single kink can be sufficient to make the progression of the dislocation easier. Indeed, as shown in Fig. 8.26, once a kink is created, the crossing of the entire dislocation can be realized by migration of the jogs towards the ends of the dislocation, by extending the length of the initial kink. In hard materials, for instance in *covalent* solids, in which the Peierls-Nabarro stress is dominant in the value of the yield strength, this effect of temperature will be very large. Thus, materials which are *fragile* at room temperature can become *ductile* at high temperature.

Finally, an increase of the temperature has the effect of lowering all the *elastic energy values* characteristic of dislocations: line tension, interaction between two dislocations, Peierls-Nabarro stress. Indeed, all these properties are proportional to the shear stress modulus μ, which decreases pronouncedly when the temperature is increased. This decrease is related to the thermal expansion of the solid which increases the distances between atoms, and, hence, decrease their binding forces (cf. section 1.2).

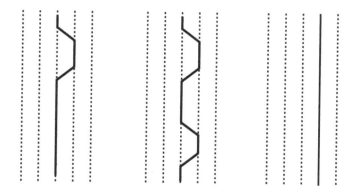

Figure 8.26   Crossing, by a dislocation (continuous line) of the distance between two Peierls valleys (dotted lines) by creation of a thermally activated "kink" surrounded by a pair of jogs.

As a consequence of all the preceding mechanisms, the *yield strength* and the *hardening* of solids are lowered. An *annealing* of a solid, i.e. its heating at high temperature followed by its slow cooling (to avoid introducing inhomogeneous internal stresses), will unblock dislocations and allow their motion and their subsequent evacuation through the surfaces of the sample. It will restore a state of the solid characterized by a low density of dislocations.

Thermal activation is essentially efficient to overcome obstacles which are of microscopic nature. Indeed, the height of the energy barriers are then of the order of the electron-volt, and one can obtain, at high temperatures a non-negligible probability of passing the barriers. Consistently, the effect of temperature is more pronounced in intrinsically "hard" materials because then, the main obstacle to the motion of dislocations is constituted by the energy barrier between the Peierls valleys and not by mesoscopic defects (precipitates, other dislocations, or clusters of point defects).

Figure 8.27 shows the variation, as a function of temperature, of the yield strength in selected solids.It appears, in agreement with the preceding argument, that copper (fcc structure) or magnesium (hcp structure) possess a weak dependence as a function of temperature. By contrast, chromium, a "hard" metal, or silicon (a covalent solid) show a pronounced dependence.

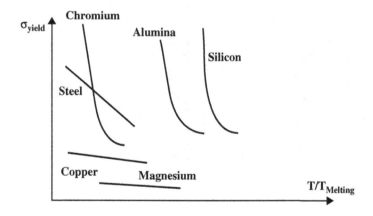

Figure 8.27 Schematic variation of the yield strength as a function of the ratio $(T/T_{melting})$ for selected materials, according to the treatise by Cottrell quoted in chapter 1. "Hard" metals, as chromium, show a large variation, while ductile materials such as copper have a small variation.

8.8 Deformation rate

8.8.1 *Orowan formula*

Consider a solid submitted to a permanent shear stress, which induces both the generation of edge dislocations (cf. section 8.4) and their propagation in their glide plane, in the direction x.

As shown in Fig. 8.5, the propagation across the sample of one dislocation of Burgers vector b induces an increment of strain (cf. Eq (8.4)) $\gamma = (b/C)$. If the average propagation-velocity of the dislocations is V and if the crystal contains a density ρ_M of mobile dislocations, all parallel to the edge A of the sample, the sample will be crossed, in the unit time, by $\rho_M.CV$ dislocations producing, each, the strain γ. The rate of increasing of the value of the shear is then (*Orowan formula*):

$$\frac{d\gamma}{dt} = \rho_M V b \qquad (8.7)$$

8.8.2 *Dislocation velocity*

The value of V in the above formula depends of the mechanisms which determine the displacement of dislocations.

In a continuous elastic medium, a dislocation is a specific strain field. In this medium, V is expected to be the velocity of propagation of a local deformation. Thus, V will be of the same order of magnitude as the velocity of sound in the solid, i.e. a few thousands meter per second. Such a value will possibly be observed at a high enough temperature, when the friction of the lattice and the interaction with various types of other objects (fixed dislocations, precipitates, etc...), does not slow the motion of the dislocation. In particular, if the propagation across the obstacles is controlled by the *climb* of dislocations it is the slow mechanism of diffusion of vacancies which will determine V.

At lower temperatures, the average velocity of dislocations is determined by the time needed to move across the obstacles. This time is controlled by the height of the energy barriers which must be overcome. Similarly to the case of the jumps of vacancies (cf. Eq.(4.15)), this is a thermally activated phenomenon (i.e. proportional to $\exp[-\Delta E/k_B T]$) which is favoured by an increase of temperature. The corresponding velocities will the be in the range $10^{-2} - 10^2 \text{cm/s}$.

Experimental observations suggest that in metals, $(d\gamma/dt)$ is often a constant as a function of time. This implies (cf. Eq. (8.7)) a constant density ρ_M reflecting a constant production of mobile dislocations to replace the ones which are evacuated from the sample. This production derives, for instance, from a generation mechanism of the Frank-Read type.

8.9   Origin of the diversity of plastic behaviours

An isolated dislocation (i.e. not interacting with obstacles of various na-
tures), submitted, on its glide plane, to a resolved shear stress larger than
the corresponding Peierls-Nabarro stress σ_{PN}, will be set into motion in
the direction of $\vec{b}$. As shown in sections 7.1.1−7.1.2, the larger the relative
width (w/b) of the dislocation-core, the smaller the value of σ_{PN}. In a
given crystal structure (w/b) is larger for the most dense planes (cf. section
7.1.1). Also note that in any structure the dislocation lines will predominantly be created in dense planes. Indeed, in these planes the interatomic
distances (hence the translation $\vec{b}$) are smallest, and consequently their
associated elastic energy, proportional to $\vec{b}^2$, will be at its minimum.

8.9.1   *Easy-glide planes; glide systems*

Table 8.2 shows the preferential glide planes (orientation and Burgers vectors) for a few simple structures. Each time, a single plane orientation
and Burgers vector have been indicated while in general several equivalent
orientations exist. For instance, in an fcc structure, as shown in section
8.3, there are four equivalent orientations for the most dense planes. In
each plane three Burgers vectors directions are equivalent. A set (plane
orientation, $\vec{b}$ direction) forms a *glide system*.

8.9.2   *Metals with an FCC or HC structure*

In agreement with the discussion in section 7.3, materials with a cubic
FCC structure (copper, gold, nickel,...) or those with a hexagonal compact
(HC) one (zinc, magnesium,...), possess the most dense atomic planes and
thus are most favourable from the standpoint of the easiness of motion
of dislocations. One therefore expects that these will generally be "soft"
metals with a low *intrinsic yield strength* (i.e. in the absence of fixed
defects). As already pointed out in section 8.6, in this case, the dominant
force opposing the motion of dislocations, which determines the *actual yield
strength*, is due to the interaction between dislocations. In this aspect, it is
relevant to distinguish the two former structures (FCC and HC).

Table 8.2 Preferential glide systems on the most dense planes of some simple structures. The plane orientation is identified by its normal vector.

Structure (example)	Glide plane	Burgers vector
fcc (Cu,Al, Ni,Au,Ag)	$(1,1,1)$	$(c/2)(1,-1,0)$
hc (Mg,Zn,Ti)	$(0,0,1)$	$(a/2)(1,\pm1,0)$
cc (Fe,Mo,Ta)	$(1,-1,0)$	$(c/2)(1,1,1)$
Diamond (C,Si,Ge)	$(1,1,1)$	$(c/2)(1,-1,0)$
NaCl	$(1,1,0)$	$(c/2)(1,-1,0)$

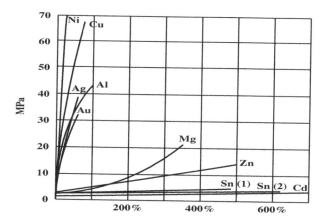

Figure 8.28   Relationship between shear strain and shear stress in metals with an FCC structure and in metals with a HC structure. The first ones are generally harder (after Schmid and Boas 1935).

Metals with the hexagonal compact structure

In this structure the most dense lattice planes are perpendicular to the hexagonal axis (Table 8.2). In these planes, the glide directions are parallel to the primitive translations of the plane's two-dimensional hexagonal lattice. The dislocation lines being located in these planes, their displacement derives from the simplest scheme described in section 8.3 and shown in Fig. 8.6: gliding in parallel planes and successive evacuation at the surface of the sample with, for each dislocation, production of an increment of shear stress. The interaction with dislocations having different directions being weak, the hardening rate is low and the process can proceed without noteworthy modifications for a very large lengthening of the sample. Such metals will therefore possess a *large ductility*.

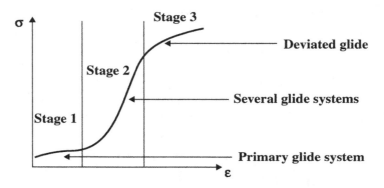

Figure 8.29   Schematic plastic behaviour of FCC metals, involving three stages in the stress-strain curve. The third stage involves cross slip.

Metals with a cubic compact structure

The situation is different from the preceding one owing to the existence (cf. Table 8.2) of several equivalent orientations of dense lattice planes (Fig. 8.3). As emphasized earlier (Fig. 8.18), a "dislocation lattice", the so-called Frank-lattice can exist, built from lines of intersection between the various dense planes. A dislocation line belonging to one orientation of dense planes is likely to intersect dislocations of the Frank-lattice, and hence be submitted to forces opposing its motion (cf. section 8.6.1). The resulting yield strength will be generally larger than in the hexagonal compact metals (cf. Fig. 8.28).

On the other hand the plastic behaviour will be more complex. It is usual to distinguish, in it, three stages.

Stage I corresponds to the motion of dislocations belonging to a single orientation of planes, constituting the *primary glide system,* for which the resolved shear stress has the largest value. This stage is similar to the behaviour described for hexagonal compact metals, with a low rate of hardening.

Stage II arises because the direction of the primary glide rotates towards the direction of the traction force (Fig. 8.2). The resolved shear relative to the other equivalent dense plane orientations increases and becomes sufficiently large to activate dislocation-sources in these planes as well as to put dislocations into motion. The simultaneous motion of dislocations in several intersecting planes will rapidly have the effect, through the interaction between dislocations, to block their motion and induce a large rate of hardening.

Stage III corresponds to a more pronounced rotation of the glide planes which has the effect of activating cross slips (cf. section 8.6.3) in other plane orientations for which the obstacles can possibly have a lower density. The dislocations can therefore "escape" in these directions and the hardening rate decreases. Figure 8.29 shows a typical stress-strain curve illustrating the former behaviour.

8.9.3 *Other solids*

Crystalline materials

In most other crystalline materials, e.g. metals with a centered cubic structure, covalent solids such as silicon or diamond, "ionic" solids such as sodium chloride, owing to the lower compacity of the glide planes, or to the strong directivity of the chemical bonds, the larger Peierls-Nabarro stress σ_{PN} is generally responsible for the value of the yield strength. Typical values as large as $10^{-2}\mu$ can then be reached.

Diamond or silicon have a very large intrinsic hardness because they cumulate two unfavourable features for the motion of dislocations: (i) a "covalent" (strongly directional) bonding of their structure, and (ii) Lattice planes having a relatively low atomic density.

Glasses and amorphous alloys

As explained in chapter 2, glasses and, more generally, "amorphous" solids (among which certain metallic alloys) are substances in which the position and orientation of the atomic groups is partly *random*. One cannot distinguish, underlying their atomic configuration any Bravais lattice or elementary unit cell. The crystallographic definition of a dislocation, which relies on the concept of a partial slip, by one lattice period, along a glide lattice plane, is therefore irrelevant. However, the "macroscopic" definition of a dislocation by the Volterra process (cf. chapter 5), assuming that these solids are continuous media, could remain valid. The actual occurence of dislocations in these materials is not experimentally attested at present. The observation method, described in chapter 5 is not applicable. Indeed, the contrast between a dislocation and its surrounding is based on the periodicity of the "good crystal" region surrounding the dislocation.

If dislocations exist in ordinary glasses, none of the conditions presiding over their easy motion would be realized. Since extended lattice planes are absent, no glide-planes are available, and, hence, any motion will have to proceed through a "climb-type" mechanism, involving the diffusion of atoms. On the other hand, the most common glasses are "silicates" whose cohesion is realized by covalent bonds Si-O similar to the ones encountered in the "diamond structure". The stress required to displace a dislocation over an interatomic distance, is likely to be very large. These reasons explain the fact that glasses are the prototype of "fragile" materials which display a complete absence of plastic range, or, in an alternate formulation, that their yield strength coincides with the stress value inducing the *fracture* of the material.

In certain amorphous metallic alloys the cohesion is realized by metallic bonds whose smaller directional character is more favourable to the motion of dislocations. Consistently, a non-negligible ductility has been observed in certain of these substances. In other metallic alloys the behaviour is analogous to that of ordinary glasses (cf. Table 8.3).

Table 8.3 Mechanical properties of some amorphous metals.

Composition	$\sigma_{fracture}$	σ_{yield}	$\epsilon_{fracture}$
$Fe_{80}B_{20}$	$\mu/10$	$\mu/10$	0
$Pd\,Si_{20}$	$\mu/20$	$\mu/30$	$< 1\%$
$Pd_{77}Cu_6\,Si_{16}$	$\mu/20$	$\mu/30$	10%

$$(8.8)$$

It is important to underline that the well known easiness of deformation of glasses at high temperatures (on which the fabrication of glass objects is based) has a different origin than the plastic deformation of crystalline metals. Indeed, it is due to the fact that glasses can be considered as "frozen-in" liquids. At high temperature a progressive "unfreezing" occurs and the glass becomes a *viscous liquid*, which is relatively easily deformable.

Quasicrystals: an intermediate behaviour

The atomic structure of *quasicrystals* (cf. section 2.4) has an underlying "lattice" built by adjoining, according to a definite rule, two types of elementary unit cells having different, but matching, shapes. The "paving" of space thus obtained has a specific regularity, distinct from that of crystals. However, similarly to the latter case, the nodes of this non-crystalline lattice can be grouped into families of parallel, *but not equally spaced*, lattice planes. Two adjacent planes are separated by several possible values of interplane spacing. Besides the sequence of distances between consecutive planes is not periodic (Fig. 8.30).

The situation has a close similarity to that of crystals. Thus, a dislocation can be defined "crystallographically" as a defect in the "normal sequence" of distances between lattice planes. Consistently, linear defects have been observed by electron microscopy and characterized in agreement with the former definition. It has been also shown that the gliding of dislocations is possible. However, the mechanisms involved are likely to be more complex than in crystals, owing to the fact that several types of energy barriers, which must be overcome during the motion of dislocations, will exist between "Peierls valleys". In practice, quasicrystalline alloys, although much more fragile than ordinary metals, possess a yield strength inferior to the fracture stress. Certain alloys can be lengthened by as much as 25% before breaking. Their elastic limit is in the range $\sigma_{yield} \approx 10^{-2}\mu$.

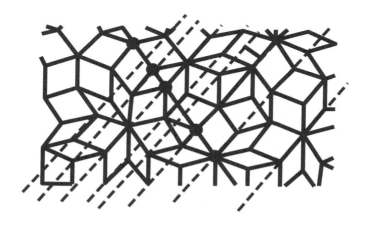

Figure 8.30 Successive non-equally spaced lattice planes in a quasicrystal.

Appendix A

Exercises

A.1 Structure of crystalline solids (chap. 2)

Exercise 1. *Stability and symmetry of an ionic crystal structure*

The stabilities of two possible structures of sodium chloride are compared.

1°) The first structure is built by replication of a cube containing 8 negative chlorine ions Cl^- at its vertices and one positive ion Na^+ at its center. The chlorine ions, of radius R, are in contact with eachother. **c** is the edge of the cubic cell.

a) Is the cubic cell a primitive cell?

b) Calculate, as a function of **c**, the total Coulomb (electrostatic) interaction energy E_s between the ions contained in a conventional cubic cell. If an ion is shared between several cells, its charge will be considered as shared between the different cells. Besides, each ion will be considered as reduced to a point-charge located at the centre of the ion. *Indication:* Calculate successively, as a function of the radius R, the negative contribution to E_s between each chlorine ion and all the sodium ions, the positive contribution between the chlorine ions, the positive contribution between sodium ions, and derive from these contributions the sum E_s. What is the sign of E_s? Can the structure be stable?

c) The cristal is built from N identical cubic cells. Explain why NE_s has the same order of magnitude as the total Coulomb energy of the crystal.

2°) The second possible structure is a close (FCC) packing of chlorine ions Cl^-, in which the sodium ions Na^+ fill all the octahedral intersticial voids of the packing.

a) Represent the configuration of ions in a single cubic cell. Assuming that the chlorine ions are in contact with eachother, calculate, as a function of R the length **c'** of the edge of the cubic cell. Is this cell a primitive cell?

b) A calculation similar to that performed in 1°) b) (and whose details are given in the solution of the exercise) show that the total Coulomb energy for a single cubic cell of this structure is:

$$E_c = -1,89\frac{e^2}{4\pi\epsilon_0 R} \tag{A.1}$$

Which of the two structures is the most stable?

Exercise 2. *Crystal structures and mechanical properties of selected solids*
Iron, copper, and carbon (in its diamond form) are solids whose crystal structures are *cubic*. ($\vec{i}$, $\vec{j}$, $\vec{k}$) are the orthogonal unit vectors along the edges of the cubic cell. $\mathbf{c}$ is the length of the edge of this cell ($\mathbf{c}_{Fe} = 2,87$Å, $\mathbf{c}_{Cu} = 3,61$Å, $\mathbf{c}_C = 3,57$Å).

1°) The crystal structure of iron is of the BCC type (body centered cubic) with one iron atom per primitive unit cell. The structure of copper is of the FCC type (face centered cubic) with one copper atom per primitive unit cell.

a) Determine as a function of $\mathbf{c}$ and of ($\vec{i}$, $\vec{j}$, $\vec{k}$), a system of *primitive* translations for the structures of iron and of copper.

b) Assuming that the atoms are in contact with eachother in each of the two structures, determine the ratio between the volume of the atoms and the volume of the crystal (*filling factor*). Which of the two structures is the most compact? The elastic stiffness of the two metals can be compared, for instance, by means of the values of their shear stiffness coefficient: $C\tilde{\;}11.10^{10}$ Pascals ($\tilde{\;}11T/mm^2$) for iron and $C\tilde{\;}7.10^{10}$ Pascals for copper. Their respective hardness can be compared by means of the values of their yield strength. For iron, it is approximately ten times larger than for copper. Does the difference of compactness of the two solids account qualitatively for their respective stiffness and hardness?

c) Dislocations are linear defects which tend to be localized in the most dense (compact) atomic planes of a crystal structure and to "glide" in the most dense directions (i.e. in which the interatomic distances are smallest) within these planes. Indicate for the two former structures the orientations of the most dense planes and, in these planes, of the most dense directions.

d) Consider in a close-packed cubic structure the succession A,B,C,... of atomic planes perpendicular to the main diagonal (111) of the cubic cell. How many closest neighbours of a given atom are situated in the same compact plane? How many are situated in each of the two adjacent parallel planes? Which figure is formed by an atom with its closest neighbours belonging to one of the two adjacent compact planes?

2°) Titanium and carbon (in its graphite form) possess hexagonal structures.

a) The structure of titanium is "hexagonal compact". Its Bravais lattice is generated by the primitive translations $(\vec{a_1}, \vec{a_2}, \vec{c})$ where $\vec{a_1}$ and $\vec{a_2}$ have the same modulus **a**. Their angle is 60°. $\vec{c}$ is orthogonal to the $\vec{a}_i$. For titanium, $\mathbf{a}_{Ti} = 2,95$Å, and $\mathbf{c}_{Ti} = |\vec{c}| = 4,68$Å. Each primitive cell contains two atoms. One is located at the origin of coordinates O and the other at O' defined by $\overrightarrow{OO'} = \frac{1}{3}(\vec{a_1} + \vec{a_2}) + \vec{c}/2$. What is the value of the ratio **c/a**, if the atoms of the structure are in contact with each other? Is this condition fulfilled for titanium? Determine the filling factor (cf. question 1°) b). Why is the result the same as for copper?

3°) The Bravais lattice of the structures of diamond and of silicon is of the FCC type. The basis is formed by two atoms respectively situated at the origin and at $(\mathbf{c}/4)(\vec{i} + \vec{j} + \vec{k})$. It is assumed that the atoms are in contact with eachother. What is the filling factor of the structure? The elastic shear stiffness coefficient of diamond is $C{\sim}50.10^{10}$Pascals and its hardness 10 to 100 times larger than that of iron. What can be said of such a result?

Exercise 3. *Quenching and annealing of vacancies in a gold crystal*

(Exercise based on two papers published in 1957 by J.S. Koehler and J.E. Bauerle in the Physical Review).

A thin gold wire (diameter $d{\sim}100\mu$, length $l_0 = 20$ cm, volume V_0) is stretched horizontally, at room temperature ($T_0 = 300$K), between two copper electrodes which inject current in the wire in order to heat it (figA.1).

It is recalled that gold crystallizes in an FCC structure. The dimension of the edge of its cubic cell is $c = 4$Å. The atomic radius of gold is $a = 1,4$Å. Its melting temperature is $T_m = 1337$K. The value of the Boltzmann constant is $k_B = 1,38.10^{-23}J/$ deg $= 0,86.10^{-4}eV/$ deg

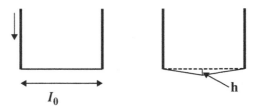

Figure A.1 (Left) Wire stretched between two electrodes injecting current. (Right) A small weight at the center of the wire allows to associate its lengthening to the height h.

1°) The wire is heated at various temperatures T comprised between 700K and 1200K and subsequently quenched (cooled down rapidly) to the room temperature. A lengthening of the wire is observed which is evidenced by placing a small weight in the middle of the wire. The wire takes the shape of a triangle of height h (Fig. A.1) which is measured by means of an optical microscope, and whose variations as a function of T are reproduced on Fig. A.2.

Explain qualitatively the observed behaviour by relating it to the presence of vacancies in the crystal structure of gold:

a) What is the origin of the lengthening?

b) Why does it increase with the maximum temperature T reached?

c) Why is the wire quenched to room temperature before measuring its lengthening?

d) Where are evacuated the atoms displaced from the vacancy sites?

2°) a) Calculate the value of the height h as a function of the atomic concentration $c = (n_{vac}/N)$ of the vacancies, of the average volume $\omega = (V_0/N)$ occupied by an atom, and of the variation of volume $\delta\Omega$ of a vacancy with respect to that of the atom previously occupying the site.

b) Deduce from figure a), the formation energy of a vacancy.

3°) a) Determine the volume variation $\delta\Omega$.

b) Specify the stress-field induced by this volume variation. Indicate the value of the corresponding elastic energy (the shear modulus of gold is $\mu \approx 2.10^{10}\text{N.m}^{-2}$)? Compare to the energy u_F of a vacancy in gold.

c) Determine the concentration of vacancies at the melting temperature.

4°) a) After heating the wire at T and then quenched it down to 40°C, it is maintained at the latter temperature. One observes that the wire regains its initial straight shape after a time interval comprised between 20 hours and 200 hours (Fig. A.2 (b)). Explain this evolution.

b) Determine the frequency of jumps of a vacancy (the oscillation frequency of gold atoms around their equilibrium position is 10^{13}Hz). Deduce from it the diffusion coefficient of vacancies (gold crystallizes in a cubic structure of the FCC type with a conventional cubic cell of size $a \approx 4$Å).

5°) a) Assume that the diffusion of vacancies occurs along a single direction (from the center of the wire to its surface, along a diameter). What are the initial and final conditions, as well as the limit conditions on the surface which must be satisfied by the vacancy concentration?

b) Show that an approximate solution of the problem is of the form $c = A\sin(\kappa z)\exp(-\kappa^2 D_l t)$. Determine A and κ. How does the concentration of vacancies evolves as a function of time?

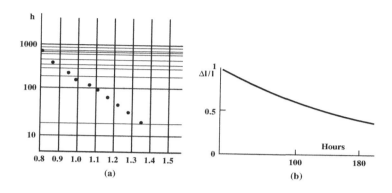

Figure A.2 (a) Value of h (expressed in microns) as a function of $(10^3/T)$. (b) Time dependence of h after quenching.

c) Deduce from the observations the value of the migration energy E_M of vacancies in gold.

6°) The wire is heated up to 800K and subsequently cooled slowly (*annealed*). One observes that it regains its straight shape. Explain the evolution which has taken place at the microscopic level?

7°) Determine the self-diffusion coefficient D_A of gold atoms at 800°C.

Exercise 4. *Doping of a semiconductor by diffusion*

1°) On the basis of Fig. A.3 (left), determine the activation energy relative to the self diffusion of silicon atoms.

2°) On the basis of Fig. A.3 (right), determine the activation energy for the diffusion of boron atoms in a silicon crystal. What can be deduced from a comparison of the two figures?

3°) A source of doping atoms (e.g. boron) is placed in contact with a crystal plate (e.g. of silicon). This source defines a dopant concentration $N_s(0)$ on the surface of the plate, which is assumed to have infinite thickness (and thus constitutes a semi-infinite medium). Show that the function $c(x,t)$:

$$c(x,t) = c_s \left[1 - \frac{2}{\sqrt{\pi}} \int_0^{\frac{x}{\sqrt{4Dt}}} e^{-\lambda^2} . d\lambda \right] \tag{A.2}$$

is a solution of the diffusion equation. How much time will be required to achieve a penetration of the boron atoms to a depth of 2 microns at 1000°C? at 1200°C?

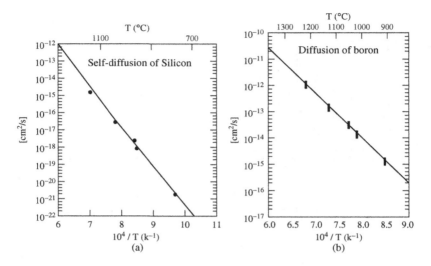

Figure A.3

The so-called "error function" $\mathrm{erf}(X) = \frac{2}{\sqrt{\pi}} \int_0^X \exp(-u^2)du$ takes the value $0,5$ for $X = 0,5$.

Exercise 5. *Energy of a vacancy and heat capacity of a solid*

Figure A.4 (a) shows the behaviour of the heat capacity $C = (dU/dT)$ of a solid gold sample as a function of temperature (U is the internal energy of the solid). In a perfect crystal, a constant value of $C = C_0$ is expected at high temperatures. However, an increase is observed in the vicinity of the melting temperature, which is assigned to the presence of vacancies. Figure 1(b) reproduces the variations, as a function of the temperature, of the logarithm $Ln[T^2 \Delta C] = Ln[T^2(C - C_0)]$. The quantity $T^2 \Delta C$ is expressed in joules.degrees^{-1}.

1°) Determine the contribution ΔU to the internal energy of the solid of the n vacancies present at the temperature T, and assumed to be without mutual interaction.

2°) Deduce from 1°) and of Fig. A.4 b) the value of the formation energy u_F of a vacancy in gold. It is recalled that the Boltzmann constant is $k_B = 0,86.10^{-4} \mathrm{eV/K} = 1,38.10^{-23} \mathrm{J/K}$.

3°) Deduce from 1°) and of Fig. A.4 (b) the approximate volume of the measured sample. Gold crystallizes in an FCC cubic structure whose conventional cubic cell has a volume equal to 64Å^3.

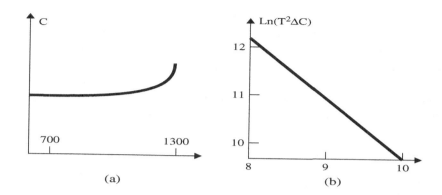

(a)

(b)

Figure A.4

A.2 Dislocations, Burgers vector (chap. 5)

Exercise 6. *Straight dislocation and plastic deformation*

A parallelipipedic crystal sample of dimensions A, B, h (height) contains a straight edge dislocation D parallel to B, and located at a distance x from the left side of the sample (Fig. A.5) Its Burgers vector is $\overrightarrow{b}$, parallel to A.

1°) Which parts of the crystal have *slipped*, due to the presence of the dislocation? What is the amplitude of their displacement? Which parts have not been affected by a displacement?

2°) The dislocation is shifted to the the right end of the sample. How is the surface of the sample modified?

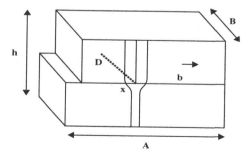

Figure A.5

3°) It is recalled that the three diagonal components $\epsilon_{11}, \epsilon_{22}, \epsilon_{33}$ of the strain tensor represent the expansion in the three directions of a local cartesian frame, and that the "shear" components $\epsilon_{12}, \epsilon_{13}, \epsilon_{23}$ represent respectively the angular variations $(1/2)\sin\theta_{12}$, $(1/2)\sin\theta_{13}$, $(1/2)\sin\theta_{23}$ between the local coordinate axes. What is the nature of the mecanical deformation induced in the ABh sample when the dislocation moves across its length A. Express, as a function of b and of h the value of this deformation.

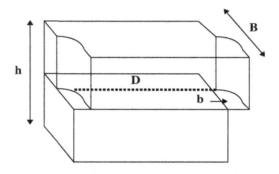

Figure A.6

4°) Determine the value of the average deformation of the sample when the dislocation is located at x.

5°) Same questions as in 1°) − 4°) if the dislocation is a screw dislocation of Burgers vector $\vec{b}$ moving parallel to B (Fig. A.6).

Exercise 7. *Rotation of a straight dislocation having a fixed point*

1°) **a)** What types of dislocations (edge, screw, or mixed) are the segments of lines CD and DE represented on the left of Fig. A.7 ? Show that they are equivalent to the insertion of an additional portion of atomic plane. Represent in a three dimensional view the deformation of the atomic planes parallel to the additional portion of plane.

b) CD is rotated around the axis DE by 45° to the back of the figure. What is then the nature of the CD dislocation ? How does it evolve if the rotation is continued? Is CD a left or right screw-dislocation when CD is in-line with AD? Check the consistency of this assignment with the standard orientation of the DE dislocation (cf. chapter 5, section 5.2).

c) Is the segment AD a dislocation?

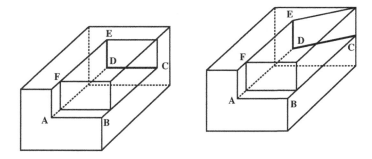

Figure A.7

Exercise 8. *Dislocations and curvature of a rod*

A crystalline metallic rod is curved (radius R) by applying an external stress. This plastic deformation is obtained by introducing a number of edge dislocations, all having the same sign (Fig. A.8). The additional half-planes associated to the dislocations permit the adjustment of the change of length of each "fiber" of the rod with respect to the neighbouring fiber which is nearer from the center of curvature of the rod. The curvature is

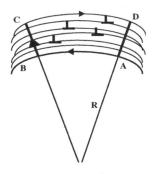

Figure A.8

assumed to be moderate (i.e. R>>AB>>b), b being the parameter of the crystal lattice. Determine, as a function of R, AB and of the thickness BC of the rod, the closure defect of the Burgers loop ABCD. Determine the number of dislocations per unit area which are necessary to account for the imposed curvature. Numerical application to a copper wire ($b \approx 2,5\text{Å}$) having a curvature radius of 10 cm.

Exercise 9. *Intersection of dislocations in an FCC structure*

a) Calculate the number of dislocations which can meet at a single node in a cubic structure of the FCC type. In this view, assume that the Burgers vectors of the various dislocations are equal to the shortest lattice translations of this structure.

b) Assume, in addition, that the dislocations are parallel to the most dense lattice rows. What is the nature of the dislocations which meet at a node?

Exercise 10. *Dislocations in a hexagonal structure*

The crystal structure of graphite is constituted by the stacking of equally spaced hexagonal carbon planes, forming a sequence ABAB... An A-plane has the carbon atoms located at the vertices of adjacent regular hexagons. A B-plane has the same structure shifted with respect to the A-planes by the translation $\vec{a}_1 = a\,\vec{i}$. Hence, half of the B atoms are above or below A-atoms. The spacing between two A-planes or between two B-planes is c. One has also $\vec{a}_2 = -(a/2)\,\vec{i} + (a\sqrt{3}/2)\,\vec{j}$.

1°) Find, as a function of $\vec{a}_1$, $\vec{a}_2$ and $\vec{c}$, a system of primitive translations ($\vec{b}_1$, $\vec{b}_2$) in each plane as well as the primitive translations of the three-dimensional structure. In each case find the number of atoms contained in the basis. What is the volume of the primitive unit cell?

2°) Determine the primitive translations of the reciprocal lattice.

3°) The dense planes of the structure are the hexagonal planes A and B. Find the possible Burgers vectors of the dislocations located in these planes. One obtains an electronic microscope image of the crystal, by using reflections of the electron beam on the hexagonal planes. Will this image provide a contrast for the dislocations? Why?

Exercice 11. *Observation of dislocations by electron microscopy*

1°) Nickel has a compact cubic structure (FCC).

a) Indicate the three sets of lattice plane orientations, defined by their normal unit vectors, which have the largest interplane spacings?

b) Denote ($\vec{a}_1$, $\vec{a}_2$, $\vec{a}_3$) the primitive translations and $\vec{c}_i$ the three perpendicular vectors defining the conventional cubic cell ($\vec{a}_i = (c/2)(\vec{j} + \vec{k})...$). Determine the primitive translations $\vec{a}_i^*$ of the reciprocal lattice as well as the vectors $\vec{c}_i^*$ defining the conventional cubic cell of the reciprocal lattice.

c) By chosing a system of primitive translations adapted to a given set of lattice planes (with two of the translations contained in the planes), show that this set can be associated to a vector $\vec{g}$ of the reciprocal lattice, normal to the set of planes, and such as $|\vec{g}| = (2\pi/d)$ where d is the distance between adjacent planes. In this view, assume that any system of primitive translations of the crystal defines the same reciprocal lattice.

2°) A thin crystal wafer of nickel is illuminated by a parallel electron beam of wavelength λ. Let $\vec{q}$ be the wavevector of the incident beam.

a) Express the condition for obtaining a reflection of the beam in a direction defined by the wavevector $\vec{q}'$.

b) Explain why the incident beam can be diffracted simultaneously in several directions $\vec{q}'$.

3°) A nickel wafer, 100 microns thick, carefully annealed for one hour at $1000°C$ is then deformed at room temperature. It is chemically thinned at its center down to a thickness of ˜1000Å. It is oriented in the microscope in order to have the electron beam parallel to $\vec{c}_3^*$. In the focal plane of the microscope, one obtains a picture reproduced in the left part of Fig. A.9, whose schematic configuration is represented on the right part.

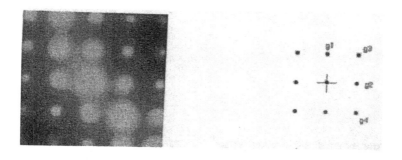

Figure A.9

a) What is the relationship between this picture and the reciprocal lattice of the crystal?

b) Identify the reciprocal lattice vectors associated to each of the spots denoted g_1, g_2, g_3, g_4, by giving their expression as function of the vectors $\vec{c}_i^*$.

4°) Four electron microscopic observations of dislocations are performed (Figs. A.10 and A.11) by isolating, each time, a single diffracted beam direction corresponding to one of the preceding spots (Courtesy of Alain Barbu, French Atomic Energy Commission).

a) Express as function of $\vec{b}$ and $\vec{G}$ the necessary condition for obtaining a contrast for a given dislocation.

b) Find from the pictures the Burgers vectors of the dislocations marked a, b, c, d, e. Is this identification possible for all the considered dislocations? What can be done to complete the observations? Whenever possible, determine the nature (e.g. edge, screw, or mixed) of the preceding dislocations.

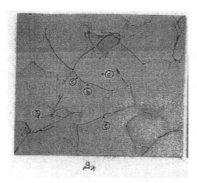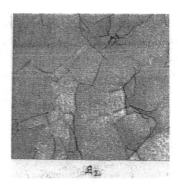

Figure A.10

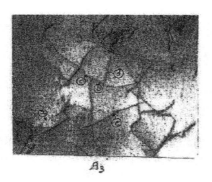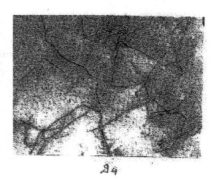

Figure A.11

A.3 Deformations and forces (chap. 6)

Exercise 12. *Forces acting on a dislocation loop*

1°) A square-loop dislocation perpendicular to z, with edges parallel to the x and y axes, is oriented opposite to the Maxwell rule with respect to the z direction. Its Burgers vector $\overrightarrow{b}$ is parallel to $x > 0$. A uniform shear stress σ_{xz} is applied to the elastic material, assumed to be isotropic, and characterized by shear constant μ. This stress tends to induce the relative gliding of the *upper part* of the material towards the direction $x > 0$ (Fig. A.12).

a) Determine the nature (edge, screw) of the various segments of the loop. Locate the additional half planes corresponding to the edge-segments. Determine the sign of the screw-segments.

b) Show, first by a qualitative argument, and then by application of the Peach-Koehler formula, that the forces applied to the loop tend to increase its surface.

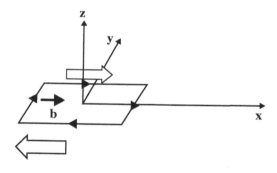

Figure A.12

Exercise 13. *Interaction between edge-dislocations having parallel glide planes*

1°) a) Recall the expression of the Cartesian components of the strain and stress tensors generated by an edge-dislocation (D_1), passing through the origin of the coordinates, parallel to $z+$ and having a Burgers vector parallel to x.

b) Give the expression of the volume expansion $\delta(r, \theta)$ associated to the strain tensor. Locate the compressed regions and the expanded ones. Locate the additional half-plane relative to the considered dislocation.

2°) A second edge-dislocation (D_2) is parallel to (D_1), and has the same Burgers vector. It passes through a point of the xy plane of polar coordinates (R, θ).

a) Find the force exerted by D_1 on D_2. Represent the components of this force, for θ belonging to each of the eight 45° sections of the plane.

b) D_1 is fixed. Assuming that D_2 can only move in the plane defined by D_2 and $\overrightarrow{b}$, what are the stable or unstable equilibrium positions of D_2? What happens if any motion is possible?

c) Same question as in b) if D_2 is an edge dislocation with Burgers vector $(-\overrightarrow{b})$.

3°) Consider a third dislocation D_{-2} having the same direction and the same Burgers vector, symmetric of D_2 with respect to D_1. The three dislocations are in the same yz plane. Assume D_2 and D_{-2} fixed. Show that D_1 is then in a stable equilibrium position (i.e. any displacement will tend to be cancelled by the forces acting on D_1).

4°) Consider a series of N edge-dislocations, all parallel, and having the same Burgers vector, aligned on the y axis, and equally spaced by a distance d.

a) Show that each dislocation, located far from the surfaces of the sample, is in stable equilibrium.

b) Show that the deformations induced by the series of dislocations add up to produce a different orientation of the two parts of the crystal, on either sides of the y-axis. Evaluate as function of b and d, the angle between the two orientations.

c) Assume that the stress field induced by each of the preceding dislo-. cations has a range limited to a cylinder of radius d around the dislocation. Calculate the elastic energy per unit area of the boundary which separates the two misoriented regions (take the core of a dislocation to be of radius b). Evaluate this energy for a desorientation $\theta = 1°$, with $\mu = 5 \, 10^{10} Pa$ (1 Pa=1J/m^3) v=(1/2) and $b = 2,5$Å.

Exercise 14. *Interaction between screw-dislocations*

In an isotropic solid, an infinite screw dislocation D, coinciding with the $\overrightarrow{Oz}$-axis, has $\overrightarrow{b}$ as Burgers vector. A second, infinite, screw dislocation D' is situated in a plane perpendicular to the $\overrightarrow{Ox}$ axis, and passes through the point $P(x = a, y = z = 0)$. Its Burgers vector is $\overrightarrow{b'}$, parallel to $(0, 1, 1)$. Show that the force between D and D' is equal to zero. In this view, denote $u = PM$ where M is the current point of the dislocation D' $(-\infty < u < +\infty)$.

Exercise 15. *Interactions between dislocations*

Two straight dislocations are orthogonal to eachother, respectively located along the axes $\overrightarrow{z'Oz}$ and $\overrightarrow{Y'AY}$, in an infinite isotropic medium (fig. A.13) refered to an orthogonal frame $Oxyz$. A is situated at $(a, 0, 0)$. $\overrightarrow{AY}$ is parallel to $\overrightarrow{Oy}$. Determine the elementary force $\overrightarrow{dF}$ exerted by the dislocation $\overrightarrow{z'Oz}$ on the element $\overrightarrow{dL}$ of the dislocation $\overrightarrow{Y'AY}$, as well as the sum of these elementary forces, and the global torque produced (sum of the moments $\overrightarrow{OM} \wedge \overrightarrow{dF}$, in which M is a point of $\overrightarrow{dL}$), in the two following cases:

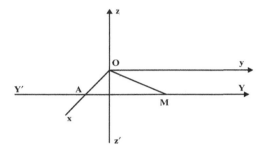

Figure A.13

1°) The two dislocations are of the screw-type and their Burgers vectors have the same algebraic value.

2°) $\overrightarrow{z'Oz}$ is a screw-dislocation and $\overrightarrow{Y'AY}$ is an edge dislocation. Both have the same Burgers vector. Show that the set of forces exerted by $\overrightarrow{z'Oz}$ on $\overrightarrow{Y'AY}$ tends to rotate the latter dislocation to bring it to an antiparallel orientation with respect to $\overrightarrow{z'Oz}$. In this final stage, what is the interaction between the two dislocations?

The following equalities can be used:

$$\int_{-\infty}^{+\infty} \frac{du}{1+u^2} = \pi \quad ; \quad \int_{-\infty}^{+\infty} \frac{u\,du}{1+u^2} = 0 \quad ; \quad \int_{0}^{+\infty} \frac{u\,du}{1+u^2} = \infty \quad (A.3)$$

Exercise 16. *Interaction of a dislocation with a planar surface*

1°) a) Recall the expressions of the displacements $\overrightarrow{\xi}$, of the strain components ϵ_{ij} and of the stress components σ_{ij}, produced, in an infinite isotropic continuous medium, by an isolated screw-dislocation passing through the origin O. Explain the sign of the non-zero components of the strain tensor.

b) A straight and infinite screw-dislocation Δ of Burgers vector $\overrightarrow{b}$, is parallel to the planar free-surface S limiting the crystal sample, at a distance l from S. Show that the mechanical state of the system (e.g. the stress field induced by the dislocation in this sample) is the same as that of an infinite medium containing Δ and a fictive dislocation Δ' (and "image" dislocation), symmetric of Δ with respect to S, and whose Burgers vector is $\overrightarrow{b}'$(fig. A.14). In this view, assume that, on a free-surface, the forces exerted on the surface by the stress field of the material are equal to zero. Determine $\overrightarrow{b}'$.

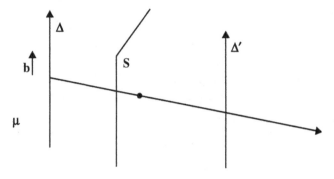

Figure A.14

c) Is the force exerted by the surface on the dislocation Δ attractive or repulsive? Calculate this force.

d) Find again the preceding result by calculating the variation of the elastic energy associated to the dislocation Δ, as a function of its distance l to the surface (The expression of the energy found, in chapter 6, for an isolated dislocation remains valid).

e) Deduce from the preceding argument the qualitative effect of the interaction of an edge-dislocation and of a free-surface parallel to the dislocation.

f) Assume that dislocations can move freely in the vicinity of the surface. Determine the direction taken by a dislocation near a free surface.

2°) Consider the planar interface S separating two isotropic elastic media having respective shear moduli μ and μ'. The first medium contains a screw-dislocation Δ of Burgers vector $\overrightarrow{b}$.

a) Show that the mechanical state, *in the medium characterized by* μ, is the same as the state induced by Δ and by an additional fictive "image" dislocation having the Burgers vector $b' = \frac{\mu'-\mu}{\mu'+\mu}b$. Show that the mechanical state *in the medium* μ' is the same as the state induced by a single dislocation Δ'' of Burgers vector $\frac{2\mu}{\mu+\mu'}b$. Assume as in 1°) b) that the continuity of the normal component of the force induced by the stress field is preserved.

b) Specify the qualitative effect of the interaction between Δ and the interface. Comment the result.

c) Deduce from b) that if S is a perfectly rigid surface (i.e. made undeformable by coating it with an oxide possessing a very high rigidity) limiting a solid having a shear modulus μ, the mechanical state of the system corresponds to the stress fields induced by the *real* dislocation Δ and by a fictive dislocation Δ', image of Δ with respect to S and having the same Burgers vector as Δ. How does Δ interact with S?

3°) A screw dislocation Δ is parallel to two free surfaces limiting an elastically isotropic solid wafer of thickness a, having a shear modulus μ (Fig. A.15).

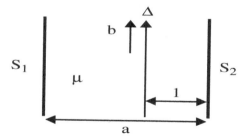

Figure A.15

a) Deduce from 1°) b) that the mechanical state of the wafer is the same as the state induced in an infinite medium by an infinite number of dislocations, successive "images" of Δ in the two surfaces S_1 and S_2. Specify the positions and signs of these dislocations.

b) Show that the stress component σ_{23} exerted on Δ by the two surfaces of the wafer can be expressed as:

$$\frac{\mu b}{2\pi} \sum_{-\infty}^{+\infty} \frac{1}{2na + 2l} \tag{A.4}$$

c) The preceding sum is equal to $(\mu b/4\pi a)\cot g(\pi l/a)$. Deduce from this expression the force acting on the dislocation.

Exercise 17. *Dislocation in a cylindrical sample*

Consider a solid, assumed to be a continuous medium (without a microscopic structure) having the shape of a cylinder of infinite height and whose section is a circle of radius R (Fig. A.16). A screw dislocation D, with Burgers vector $\overrightarrow{b}$ is parallel to the axis of the cylinder. The surface of the cylinder is "free": no force is exerted on it by the stress field present in the solid (or alternately, the stress field vanishes on the surface).

1°) The dislocation is located at the center of the cylinder (Fig. A.16 (a)). Show, by examining its effect on the surface of the cylinder, that the stress field is the same as that induced in an infinitely extended solid.

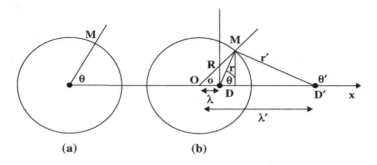

(a)   (b)

Figure A.16

2°) The dislocation is located at a distance $\lambda = OD$ from the center, along the x-axis (Fig. A.16 (b)). Express, at a point M of the surface, as a function of R, λ, and of the polar angle $\phi = \widehat{xOM}$, the non-zero components σ_{ij} of the stress tensor *which would be induced in an infinite solid* by the dislocation D. Note that one can write:

$$r \sin \theta = R \sin \phi \; ; \; r \cos \theta = R \cos \phi - \lambda$$

3°) Determine the force exerted by these stress components on an infinitesimal portion of the surface surrounding M, and normal to the vector $\overrightarrow{n} = (\cos \phi, \sin \phi)$. Does the result of question 1°) remain valid?

4°) On the basis of the results of 2°), show that the stress field induced by D inside the cylinder is the same as the sum of the fields induced in an infinite solid by D and by an additional fictive dislocation D', parallel to D, having the Burgers vector $\overrightarrow{b'} = -\overrightarrow{b}$. Determine the coordinate λ' on the x-axis, of the dislocation D'. Two values will be found, one of which only is physically acceptable.

5° a) Is the plane (D, Ox) a glide plane for D?

b) Determine the force exerted on D by the surface of the cylinder. What will be the effect of this force, if D is mobile in its glide plane?

c) Will there be a jog on the surface of the cylinder if D is evacuated?

A.4   Core of a dislocation (chap. 7)

Exercise 18. *Core energy of a dislocation*

It is recalled (Eq. (7.6) in chapter 7) that the increase of the "elastic part" of the energy of the core of an edge dislocation is, for a length b of the dislocation (i.e. the interatomic spacing):

$$K\sum\{\xi(x_{n+1}^0) - \xi(x_n^0)\}^2 \qquad (A.5)$$

On the other hand, the shear stress which corresponds to the misorientation of the bonds on either sides of the glide plane is (cf. Eq.(7.12)):

$$\sigma = \pm\frac{\mu b}{2\pi a}\sin(\frac{4\pi\xi(x_n^0)}{b}) \qquad (A.6)$$

As shown in chapter 7, the function $\xi(x)$ and its derivative $\xi'(x)$ which describe the atomic displacements within the core of a dislocation are such as:

$$\xi(x) = -\frac{b}{\pi}\arctan[th(\frac{2x}{w})] \qquad (\frac{d\xi}{dx})^2 = \frac{\mu b^2}{2\pi^2 aK}\cos^2(\frac{2\pi x}{b}) \qquad (A.7)$$

Assume, for the sake of simplicity that $a = b$.

1°) Show that the total energy of the core can be expressed as:

$$U_c = \int_{-\infty}^{+\infty}[Kb(\frac{d\xi}{dx})^2 + \frac{\mu b^2}{8\pi^2}\sin^2\frac{4\pi\xi}{b}]dx \qquad (A.8)$$

2°) Show that U_c is equal to:

$$U_c = \frac{\mu b^2 w}{4\pi^2}I \qquad (A.9)$$

in which I is a dimensionless integral whose value will be assumed to be smaller than 2.

3°) Compare the energy of the dislocation-core to the elastic energy induced by the dislocation in the "good" crystal (the sample linear dimension will be taken equal to ~100 microns.

Exercise 19. *Dissociation of the dislocation core*

1°) In a cubic FCC monoatomic structure, let A and B be two consecutive atomic planes of the stacking of atomic planes CAB-CAB...perpendicular to the direction (111) (Fig. A.17).

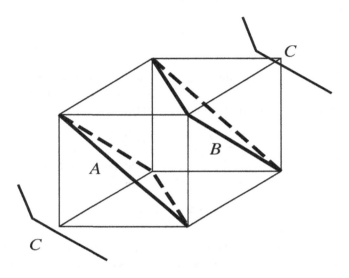

Figure A.17

Figure A.18 represents, projected on the (111) direction, the atomic configuration of the two planes in which an "imperfect" dislocation is present. The atoms of the A-plane (represented shade in grey, and situated at the back of the figure) form a regular triangular lattice. The atoms of the B-plane are represented in black or in white and are situated in front of the figure. In the lower part of the figure (white atoms) they are located at the center of every second A-triangle. These positions are assumed to be the reference positions of a normal B-plane. In the upper part of the figure (Black atoms) are at the centers of the empty triangles of the B-structure. These positions are the standard positions of a C-plane. They are related to the standard positions in a B-plane by a partial slip of the upper part of the plane, and thus generate a dislocation line D (horizontal line within the B-plane represented on Fig. A.18). On either sides of D, in its vicinity, the atoms of the B-plane occupy intermediate positions between the standard positions in a B-plane and the standard positions in a C-plane.

1°) a) $\vec{c}_i$ (i=1,2,3) are the edges of the conventional cubic cell. Express, as functions of the $\vec{c}_i$, the primitive translations joining the atoms in the A-plane and in the undeformed structure of the B-plane.

b) Express, as function of the $\vec{c}_i$, the Burgers vector $\vec{b}$, represented in Fig. A.18, which relates the atomic positions on either sides of the core of the dislocation D. Determine the Burgers vectors of the edge-component and of the screw component of D.

c) Is $\vec{b}$ a vector of the Bravais lattice? Is D an ordinary (perfect) dislocation? Which type of defect is limited by D.

2°) A second line D' of discontinuity is created in the B-plane (Fig. A.19), parallel to D, associated to a slip of the C-structure, of amplitude $\vec{b}' = (1/6)(\vec{c}_2 + \vec{c}_3 - 2\vec{c}_1)$.

a) Is D' a perfect dislocation? Determine the Burgers vectors of its edge and screw components. What is the nature of the atomic positions in the portion of the B-plane situated beyond D'?

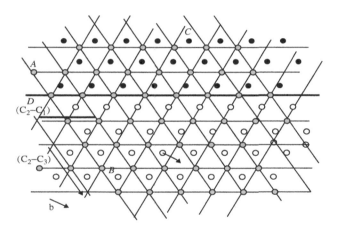

Figure A.18

b) Which type of defect is obtained if the two lines D and D' are brought to coincide?

3°) a) Coming back to the situation of separation, at a distance d, of the two dislocations D and D', determine the force exerted by one dislocation on a unit length of the other. The Poisson coefficient ν of the material (cf. chapter 3) will be taken equal to 0.5.

b) The band of stacking fault comprised between D and D' is assumed to increase the energy of the material proportionally to the area of the band by an amount γ per unit area. Determine the effective force between D and D', related to the stacking fault.

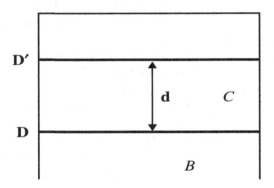

Figure A.19

c) Show that an equilibrium spacing (d_{eq}) exists between D and D'.

d) Using an electron microscope, an observation shows that, in copper ($c \approx 3.6$Å; $\mu \approx 7.5.10^{10}$J/m^3), the value of d_{eq} is 50Å. Deduce from this value the energy of the stacking fault in this metal, for a length of one micron of the dislocations.

e) It is assumed that, at equilibrium, the interaction energy between D and D' is twice the energy of the part related to the stacking fault. Determine this energy. Compare it to the elastic energy induced in the solid, in a cylinder of radius 1μ, by D and D'.

4°) Explain why a perfect dislocation tends to dissociate into two imperfect dislocations D and D'.

Exercise 20. *Dissociation of a dislocation*

A crystalline solid has a cubic structure with conventional cell $(c\overrightarrow{i}, c\overrightarrow{j}, c\overrightarrow{k})$. It is recalled that a "perfect" dislocation has a Burgers vector equal to a vector of the Bravais lattice.

1°) Consider, first, that the Bravais lattice is body centered cubic (BCC).

Show that a perfect dislocation with Burgers vector $\overrightarrow{b} = c(\overrightarrow{i} + \overrightarrow{j})$ is less stable than a set of two parallel, perfect dislocations, with, respectively, Burgers vectors $\overrightarrow{b}_1$ and $\overrightarrow{b}_2$, with $\overrightarrow{b}_1 = (c/2)(\overrightarrow{i} + \overrightarrow{j} + \overrightarrow{k})$. The two dislocations will be assumed to be sufficiently distant from each other in order that their interaction can be neglected.

2°) Consider now a solid with a face centered cubic (FCC) structure, and a straight, perfect dislocation D, situated in a compact plane perpendicular to the (111) direction, with Burgers vector $\overrightarrow{b}' = (c/2)(\overrightarrow{i} - \overrightarrow{j})$.

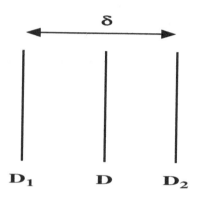

Figure A.20

a) Show that D tends to dissociate into two dislocations D_1 and D_2 (Fig. A.20), whose interaction will be neglected, having respective Burgers vectors $\overrightarrow{b}'_1$ and $\overrightarrow{b}'_2$ with:

$$\overrightarrow{b}'_1 = \frac{c}{6}(2\overrightarrow{i} - \overrightarrow{j} - \overrightarrow{k})$$

b) The interaction between D_1 and D_2 is now taken into account. Assume that D_1 and D_2 are two screw dislocations, parallel to D. D_1 has the Burgers vector $\overrightarrow{b}'_1$ and D_2 has, as Burgers vector, the projection of $\overrightarrow{b}'_2$ on $\overrightarrow{b}'_1$. Determine the force exerted by one dislocation on the unit length of the other, when their spacing is δ. Is the interaction attractive or repulsive? Could this conlusion be anticipated?

c) Are D_1 and D_2 perfect dislocations? What can be said of the portion P of plane, of width δ separating D_1 and D_2?

d) P induces an increase of energy f per unit area. Show that this increase corresponds to an attractive force between D_1 and D_2, whose value per unit length is equal to f. Deduce from this result the value δ_0 of thequilibrium spacing between the two dislocations.

e) Assume that δ_0 is the "effective width" of the core of the initial dislocation D. Determine the minimum value σ of the stress needed to set into motion this dislocation within the compact plane (111), in a direction parallel to its Burgers vector. Numerical application: In copper, $c = 3, 6$Å, $\mu \approx 7.10^{10}$J/m^3, $f \approx 10^{-1}$J/m^2. Calculate (δ_0/b') and σ. Compare the latter value to the values of the yield stress currently observed in metals.

A.5   Mechanism of plasticity (chap. 8)

Exercise 21. *Deformation of a metal rod*

A monoatomic metal has a hexagonal compact structure having primitive translations $(\vec{d}_1, \vec{d}_2, \vec{c})$, with $|\vec{d}_1| = |\vec{d}_2| = 2,5$Å ; $(\widehat{\vec{d}_1, \vec{d}_2}) = \pi/3$; $\vec{d}_i.\vec{c} = 0$. A traction is exerted along an axis located within the $(\vec{d}_1, \vec{c})$ plane, and making an angle $\phi \neq 0$ with $\vec{c}$. The resulting lengthening, along this direction, is the cumulative effect of glides as represented on Fig. 8.2 of chapter 8.

1°) a) Along which lattice planes of this structure, and along which directions in these planes, are the glides easiest?

2°) Assume that a single of these directions, denoted $\vec{g}$, is activated as a glide direction. A rod of the metal has been annealed and contains *in the glide planes*, prior to any deformation, dislocations all parallel to eachother and which can glide in the $\vec{g}$ direction. Their density is equal to $10^4.cm^{-2}$. The dislocations (denoted D) are uniformly distributed in the volume of the sample. The rod also contains other dislocations having directions approximately perpendicular to the preceding glide planes (forest of dislocations).

a) The cylindrical rod has a diameter equal to one centimeter. How many dislocations D can be found in each possible glide plane parallel to the considered plane direction. Determine the average spacing between planes containing a dislocation.

b) Among all the planes parallel to the considered plane direction, determine the planes which will effectively glide. Can this result explain the observation of a spacing equal to $\approx 10^4$Å between glide planes?

c) Which mechanism can explain the magnitude of the cumulated glide amplitude: $\approx 10^3$Å?

Exercise 22. *Value of the yield stress*

The structure of each of the metals shown in the table hereunder, is formed by a stacking of atomic planes, along the c-axis, according to a sequence ABAB... . Each plane has a hexagonal-compact structure. However the three-dimensional structure differs from the ideal hexagonal compact structure by the fact that the ratio (c/a) of the primitive translations normal and parallel to the hexagonal planes differs from the standard value $2\sqrt{2/3}$. Neglecting the directional character of the bonds between atoms, and assuming that the yield stress of the material is determined entirely by the value of the Peierls-Nabarro stress of the hexagonal-compact planes, classify the three metals listed by increasing values of their yield stress. Justify this classification.

métal→	Beryllium	Cobalt	Zinc
c/a	1,56	1,62	1,86

(A.10)

Exercise 23. *Yield stress*

1°) The curve OABC in Fig. A.21 (a) represents the mechanical behaviour of a solid sample (e.g. a thin metallic "whisker") which does not contain *initially* any dislocation. Explain the shape of this curve. Determine the order of magnitude of the stresses σ_A and σ_B as functions of the shear modulus μ and of the characteristics of the dislocations in this material.

2°) An edge-dislocation D with Burgers vector $\overrightarrow{b}$ (Fig. A.21 (b)) moves to the right, in its glide plane, when a shear stress σ is exerted. The glide plane contains a set of small spherical precipitates of radius $r \gg W$, distributed uniformly as shown on the figure, with $A \gg a$. Whenever D approaches a vertical row of precipitates, each portion of D close to a precipitate is stopped at the left of the precipitate which acts as a pinning center. The remaining length of D continues to glide by deforming its shape.

a) Represent on a drawing the successive shapes of D when the value of the applied stress σ increases.

b) Assume that the Peierls-Nabarro stress of D in its glide plane can be neglected. Determine as function of μ, b, W, a and A, the value of the yield stress σ_L of the material. The stress field induced by D in the material will be assumed to extend to an approximate distance $\approx A$.

c) What remains behind the dislocation after the continuation of its motion across the row of defects?

d) Assuming that a dislocation-loop of radius r only deforms the surrounding material at a distance $\tilde{\ }r$, determine the *order of magnitude* of the increase of elastic energy of each precipitate after interaction with a moving dislocation D.

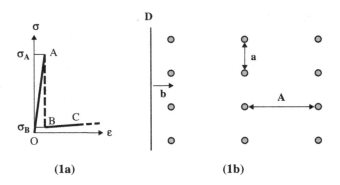

(1a)   **(1b)**

Figure A.21

Exercice 24. *Reaction between dislocations*

Two oriented straight dislocations $\overrightarrow{AB}$ and $\overrightarrow{CD}$ are respectively situated in a vertical plane and in a horizontal plane whose intersection is the y axis (left part of Fig. A.22). $\overrightarrow{AB}$ is parallel to $(0,-1,-1)$ and $\overrightarrow{CD}$ is parallel to $(1,1,0)$. With the former choice of orientations, their Burgers vectors are both equal to $\overrightarrow{b} = (0b0)$.

1°) Specify the nature of the two dislocations and their respective glide planes.

2°) a) Express their line tensions T_{AB} and T_{CD}, for an approximate linear size of the sample equal to R, and a core radius equal to r_C. The shear modulus of the isotropic material is μ and the Poisson coefficient ν.

b) What is the effect of the line tension of a dislocation of arbitrary shape on the length and shape of the dislocation?

3°) A shear stress $\sigma_{yz} = \sigma > 0$ is applied, tending to induce a slip of the upper part of the sample, situated above the horizontal plane, with respect to the lower part, parallel to Py. Determine the force exerted on each of the two dislocations. Refer its direction to the glide plane. Specify its orientation.

4°) σ is assumed to be larger than the Peierls-Nabarro stress in each of the glide planes. Show that, at room temperature, σ is not sufficient to set in motion AB, while CD is displaced by such an amount as to intersect AB at point P of the y-axis (right part of Fig. A.22).

5°) a) When APB and CPD intersect at P, is it consistent to consider CPB and APD as dislocations (each one being formed by two segments)?

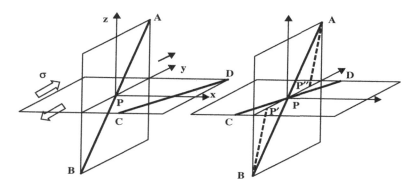

Figure A.22

b) By relying on the result of 2°), explain why, in the absence of shear stress ($\sigma_{yz} = 0$), the two dislocations will tend to separate and give rise to two dislocations $CP\prime B$ and $AP"D$. $P\prime$ and $P"$ originate from the splitting of P in opposite senses along the y-axis (Fig. A.22, dashed lines). Which energy is thus lowered? Are the motions of the different segments CP, PB, AP, PD, which allow this separation, possible at room temperature?

6°) Assume that the positions of the two dislocations $CP\prime B$ and $AP"D$ are stabilized, and that a shear stress σ_{yz} is, again, applied.

a) Show that σ_{yz} exerts a "climb" force on the portions $P\prime B$ and $AP"$ and a "glide" force on $CP\prime$ and $P"D$.

b) Determine the segments which cannot move at room temperature. Describe the shapes which will be taken by $CP\prime$ and $P"D$ during their motion which is induced by σ_{yz}. What type of dislocation source will have thus been created?

c) In which plane these sources will generate glides and thus allow the plastic shear of the sample?

Exercise 25. *Dislocation source*

Let $\overrightarrow{AB}$ be a straight portion of an edge dislocation whose ends are pinned to the lattice. The Burgers vector $\overrightarrow{b}$ is normal to the plane of the Fig. A.23, and directed towards the observer. The orientation of the dislocation is assumed to be conform to the standard convention defined in chapter 5.

a) Specify the orientation of the additional half-plane associated to the edge-dislocation.

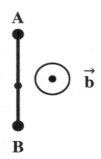

Figure A.23

b) What type of stress must be applied in order to induce a curvature of the *AB* dislocation to the right, within the plane of the figure?

c) Is the preceding motion of the dislocation a glide or a climb? Which condition must be realized to achieve such a motion?

d) Assuming that the conditions of such a motion are fulfilled, and that the source can then work similarly to a Frank-Read source, which modification of the solid occurs at each cycle of the source? Is this modification consistent with the direction of the applied stress?

Exercise 26. *Dual Frank-Read "windmill"*

Consider two parallel segments of dislocations parallel to the *y*-axis. Their Burgers vectors $\vec{b}$ et $\vec{b}\,'$ have the same modulus, the same direction parallel to the *x*-axis. This set of vectors can have three possible configurations represented on Fig. A.24. Each segment has its ends pinned as in a Frank-Read source.

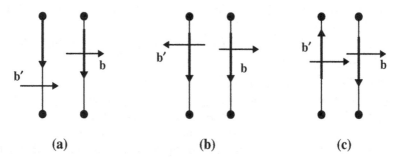

(a)   (b)   (c)

Figure A.24

A uniform shear stress σ_{xz} is applied.

1°) The interaction between the dislocations is neglected at first. Describe the sequence of shapes of the segments when the magnitude of σ_{xz} increases. Can this system generate a succession of dislocations similarly to the Frank-Read source?

2°) The interactions are now taken into account. Which of the three configurations will correspond to the activation of the dual source by the smallest value of σ_{xz}?

Exercise 27. *Dislocations energy*

A cubic sample of copper has edges parallel to (x, y, z), one centimeter long each. It is strongly hardened by application of a shear stress σ_{xz}. During this process, an amount of 100 joules of mechanical energy is stored in the sample through an increase of the number of dislocations.

a) Explain why all the dislocations are approximately parallel to echother at the end of the hardening process.

b) Calculate the order of magnitude of the dislocation density (number per cm^2) at the end of the hardening process. The elastic shear modulus of copper is $\mu \approx 7.10^{10} \mathrm{J/m}^3$ and the Burgers vector of the dislocations is $b \approx 2,5 \mathrm{\AA}$.

Exercise 28. *Edge dislocation and vacancies*

1°) A monoatomic crystalline solid has a simple cubic structure (SC) with a cubic cell of edge b. It is free from any internal or external stress. Let u_F be the formation energy of a vacancy. Determine, assuming thermodynamic equilibrium at the temperature T, the probability q_0 that an atomic site be "occupied" by a vacancy.

2°) The volume of the solid in the region of the vacancy is submitted to a stress $\overline{\overline{\sigma}}$ which induces, in particular, a local pressure $p = -(1/3) \sum \sigma_{ii}$.

a) The presence of the vacancy is associated to a volume *contraction* $\delta\Omega < 0$ with respect to the situation in which an atom was present. Explain why the energy of formation of the vacancy is *decreased* (by δu_F) if $p > 0$, and *increased* by the same amount if $p < 0$.

b) Evaluate δu_F. Deduce from this value the probability q that an atomic site be occupied by a vacancy in a region of the solid submitted to a pressure p.

3°) The solid contains a straight edge dislocation passing through O, parallel to $\overrightarrow{Oz}$, and having the Burgers vector $\overrightarrow{b}$ parallel to $\overrightarrow{Ox}$. The $\overrightarrow{Oy}$-axis is directed towards the additional half plane associated to the dislocation.

a) Explain why the dislocation induces, at equilibrium, a non-uniform spatial distribution of vacancies in its neighbourhood.

b) Deduce from 2°) b) and from the characteristics of the fields induced by an edge-dislocation, the probability $q(r, \theta)$ of occupation of an atomic site by a vacancy in the vicinity of the dislocation. (r, θ) are the polar coordinates of a site referred to the axes (O, x, y, z). Specify the region in which an excess of vacancies can be found with respect to the situation in the absence of dislocation.

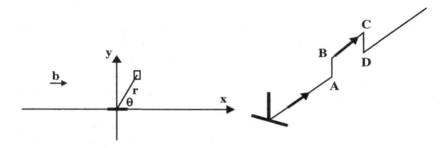

Figure A.25

c) Calculate $q(r = 3b, \theta = +\pi/2)/q_0$ for $\nu = (1/2)$ (Poisson coefficient), $\delta\Omega = -b^3/4$, $T = 1200K$. The following values will be taken: $\mu b^3 \,\tilde{}\, 10\text{eV}$, $k_B.1200K \,\tilde{}\, 10^{-1}\text{eV}$. If $u_F = 1\text{eV}$, calculate $q(r = 3b, \theta = +\pi/2)$.

4°) Assume that the preceding q-value is the same in the entire volume of the dislocation core.

a) Explain why the shape, at the atomic scale, of the dislocation line involves "jogs" similar to AB and CD as shown in Fig. A.25. Specify the amplitude of these jogs.

b) Deduce from a) that the vacancies induce an *effective motion of the dislocation* parallel to y, of amplitude qb. Characterize this type of motion of a dislocation.

c) A compression σ_{xx} is applied to the solid. Detremine the direction and sense of the forces exerted on the dislocation line and on the jogs AB and CD.

Exercise 29. *Climb of a dislocation*

This exercise is aimed at calculating the *climb velocity*, at a temperature T, of an edge-dislocation D, in a material submitted to a uniform compression stress $\sigma_{xx} = -\sigma < 0$ (Fig. A.26 (left)). All the interatomic distances in the directions x, y, and z will be taken equal to b.

1°) Determine the direction and sense of the force exerted by σ on D. Determine its value per unit length of D. Can this force move the dislocation at room temperature? Why?

2°) The formation energy of a vacancy in the material is u_F. Its migration energy is E_m. Its presence (i.e. the absence of the corresponding atom) induces a local variation of volume $(-\Delta V) < 0$. The corresponding *decrease* of energy in presence of a local hydrostatic pressure P is equal to $P\Delta V$.

a) Recall the expression of the non-zero components of the stress tensor $\sigma_{ij}(r, \theta)$ induced by D in the solid. Deduce the energy variation $(-\Delta u)$ associated to a vacancy, due to the stress field, as a function the position of the vacancy with respect to D.

b) At some distance r of the dislocation core $(r \geqslant R_0)$, the concentration of vacancies is the thermal equilibrium concentration at T. Show that the maximum atomic concentration $(c = n/N)$ of vacancies is reached at the minimum distance R_0 of D. Determine $c_{\max}$, as function of $u_F, \Delta V, \mu, \nu, b$, and R_0? The effect of the external stress σ will be neglected.

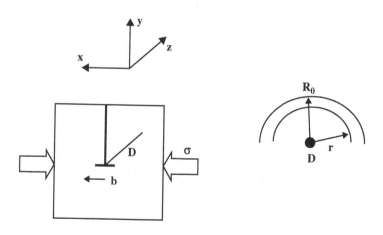

Figure A.26

3°) If a vacancy reaches the dislocation core, it is "absorbed" by D and disappears. Which modification is induced in the dislocation by this absorption? Same question relative to the absorption of a large number of vacancies.

4°) Consider the diffusion of vacancies between the region of the material in which the concentration of vacancies has maximum value ($r = R_0$), and the core of the dislocation. In this view it is assumed, for the sake of simplicity (Fig. A.26 (right)) that:

- The concentration of vacancies is uniform at a given distance of D ($c = c(r)$).

- The one-dimensional diffusion equation (coordinate r) can be used.

- A steady state is established ($\partial c/\partial t = 0$).

a) Show that in the region between D (origin of the y-axis) and R_0 the concentration in *volume* is:

$$c(r) = \frac{c_{\max}}{b^3} \frac{r}{R_0} \tag{A.11}$$

b) Determine the flow J of vacancies (number of vacancies per unit area and per second) which reaches the dislocation. Determine the number of vacancies per second which reaches a length b of the dislocation.

c) Deduce from b) the climb velocity **v** of the dislocation expressed in terms of the number of interatomic distances per second.

d) Evaluate numerically the velocity for $u_F = 1$ eV, $k_B T = 0.1$ eV, $D_L = 2.10^9.b^2, \nu = 0.5, R_0 = 10b, \mu\Delta V = 0.2$ eV.

Exercise 30. *Nucleation and growth of a dislocation loop*

Consider in an elastically isotropic infinite solid (Fig. A.27) a planar circular dislocation loop of center O, of radius ρ and of Burgers vector $\overrightarrow{b}$ parallel to the plane of the loop.

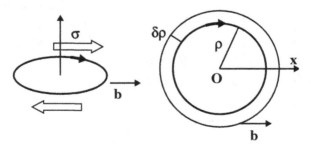

Figure A.27

a) Explain qualitatively why the stress field $\overline{\overline{\sigma}}^L$ induced by the loop in the solid vanishes at distances $r \gg \rho$ of the center of the loop. It will be assumed in the following questions that $\overline{\overline{\sigma}}^L$ is non-zero in a torus of radius ρ centered on the loop.

b) Specify the nature of the different elementary dislocation portions of the loop (edge-, screw-, or mixed-).

c) Let E_e be the elastic energy associated to the stress field $\overline{\overline{\sigma}}^L$. Assume that, on each elementary portion of the loop, the elastic energy per unit length has the same expression as for a straight dislocation. Show then that:

$$E_e = \frac{\mu b^2}{4\pi} \frac{2-\nu}{(1-\nu)} \pi\rho Ln(\frac{\rho}{r_0}) \qquad (A.12)$$

d) The loop is submitted to a uniform shear stress σ (Fig. A.27 (left)). Let $\delta\rho$ be a spontaneous fluctuation of the radius ρ tending to increase the area of the loop (Fig. A.27 (right)). Evaluate the increase of elastic energy δE_e of the solid. Evaluate the work δW done by σ and associated to $\delta\rho$.

e) By studying the sign of $\delta U = (\delta E_e - \delta W)$ explain why the loop will tend to reduce spontaneously its size and disappear if $\rho < \rho_C$, in which the "critical" radius ρ_C is determined by the equation:

$$\rho_C = \frac{\mu b}{8\pi\sigma} \frac{2-\nu}{(1-\nu)}[1 + Ln(\frac{\rho_C}{r_0})] \qquad (A.13)$$

Calculate approximately ρ_C as function of b for $\sigma = 10^{-2}\mu$, $\nu = 0.5$, $r_0 = 3b$ (Two solutions will be found, one of which only, $\rho_C < r_0$, is physically acceptable).

f) Evaluate the amount of "thermal energy" U (expressed in electron-Volts) which must be available in the solid, submitted to the stress σ, to create a loop of radius ρ_C. Make an appropriate comment. The value $\mu b^3 \approx 4eV$ will be used.

Exercise 31. *Plasticity of a polycrystalline material* A metallic poly-crystalline material is formed by grains of diameter d ˜100μ. The compression and shear elasticity moduli are respectively $C \approx 2.10^{11}N/m^2$ and $\mu \approx 10^{11}N/m^2$. The value of the (dimensionless) Poisson coefficient is 0.4. All the interatomic distances will be taken equal to $b \approx 2.5$Å.

1°) Determine the relative value (W/b) of the width of the dislocation core (assume that $K = Cb$). Deduce from the result the value of the Peierls-Nabarro shear stress σ_{PN} needed to set in motion a dislocation within a dense plane (Use the value $\gamma = \pi$ in formula (7.23)).

2°) a) A Frank-Read source of edge dislocations having a 10 μ length, is located in a dense plane in the center of a grain within the sample. Determine the value of the minimum shear stress σ_{FR} required to activate the source.

b) The reduced shear stress acting on the dense planes of a grain results from a traction σ_T applied to the sample. The direction of the traction-force makes angles $\theta \approx \phi \approx 45°$ with, respectively, the slip-direction, and with the glide-plane within the grain. Determine the value of σ_T which corresponds to σ_{FR}.

3°) A dislocation produced by the Frank-Read source, and moving towards the right end of the grain, is blocked by the grain boundary separating the considered grain from the neighbouring one.

a) Explain why all the dislocations produced by the source have the same glide plane.

b) Determine the force F_σ per unit length exerted by σ on each dislocation. Specify its direction and sense.

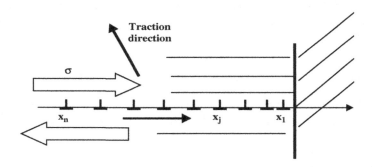

Figure A.28

c) The grain boundary blocks the motion of the "first" dislocation which moves ahead of the n dislocations produced by the source. Explain why the motion of the $(n-1)$ other dislocations is also blocked.

d) Why can the activity of the source be also stopped? Assume that the n dislocations are all far from the source and close to the grain boundary. For which value of n will the source be disactivated? Determine the value of n for $\sigma = 2\sigma_{FR}$.

4°) a) Within the series of dislocations blocked by the grain boundary, each dislocation except the "head" one is in equilibrium under the effect of the interaction with the other dislocatoons and of the force exerted by applied shear stress σ. Write the equation expressing this equilibrium for a dislocation located at the coordinate x_p (the x-axis is oriented to the right of the figure).

b) The interaction energy of the n dislocations is of the form $W = W(x_1 - x_2, x_1 - x_3, ..., x_p - x_{p'}, ...)$. The force exerted on one of the dislocations is $F_j = -\partial W/\partial x_j$. Determine the force exerted on the "head" dislocation by the $(n - 1)$ other dislocations. Deduce from this result that the grain boundary exerts the force $f = n\sigma b$ on the head dislocation.

c) Show that the result $f = n\sigma b$ can also be found by assuming that the n dislocations are rigidly linked to each other, and by evaluating the work effected by the stress σ when the n dislocations are globally displaced along x.

d) The grain boundary is assumed to exert a force F_G on the head dislocation. How many dislocations must pile up in front of the grain boundary in order to "push" the head dislocation across the boundary and resume its motion ($F_G = 20\sigma_{FR}.b$). Describe the mechanism by which the successive dislocations produced by the source will move across the sample.

5°) Assume that the n dislocations are equally spaced. Deduce from the equilibrium equation written in 4°)a) the length of a series of 10 dislocations submitted to the shear stress σ (Use the formula $\sum_1^9 \frac{1}{p} \approx 2, 8$). Determine this length for $\sigma = 2\sigma_{FR}$.

6°) Let F_G be the force exerted by the grain boundary on the head dislocation. Determine the minimum value of the shear stress σ, required to unblock the motion of the dislocations in the sample, as a function of the linear size d of the grain (use the approximation $\sum_1^{n-1}(1/i) \approx 3$). Numerical value of σ for $d = 10$ microns.

7°) A "virgin" sample " has a density of mobile dislocations equal to $10^4/cm^2$. The minimum observable deformation is 10^{-5}. Assuming that the mobile dislocations move freely across the grain boundaries, is the yield stress equal to σ_{PN} or σ_{FR}? To determine the answer, consider successively the deformation of a single grain and of an entire polycrystalline sample of area 1x1 cm^2 perpendicular to the mobile dislocations.

Exercise 32. *Polygonization of a rod*

Consider as in exercise N°8 the curvature of a metallic rod (radius R) through application of an external stress. Structurally, the curvature results from the introduction of edge dislocations of identical signs. Assume that the dislocations can only move freely in their glide plane. Referring to the results of exercise N°13, explain why the rod will eventually take the shape of a polygon instead of keeping its circular shape.

Appendix B

Solutions to Exercises

Exercise 1. *Stability and symmetry of an ionic crystal structure*

a) It is a primitive unit cell: The shortest translations generating the structure are the edges of the cubic cell.

b) In order to evaluate the Coulomb interactions within a cell considered as a neutral object, a charge (-1/8) must be assigned to the chlorine ions located at the vertices of the cube. The sodium ion at the center, which is not shared between several cells has the charge (+1). Putting $c = 2R$, the interaction energy

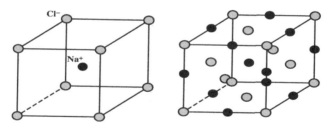

Figure B.1

between sodium ions and chlorine ions is:

$$\sum \frac{qq'}{4\pi\epsilon_0} \cdot \frac{1}{d} = \frac{(-e^2/8)}{4\pi\epsilon_0} [8. \frac{1}{c\sqrt{3}/2}] = \frac{(-e^2)}{4\pi\epsilon_0} \frac{1}{R\sqrt{3}} = -0,577 \frac{e^2}{4\pi\epsilon_0 R} \quad (B.1)$$

On the other hand, there are three contributions to the positive interaction between chlorine ions: interaction between neighbouring ions (distance c) , interaction between opposite ions on a face diagonal, interaction between ions separated by a cube diagonal. This has to be taken into account for each chlorine ion and the sum must be divided by 2 to take the double counting into account.

$$\frac{8}{2} \cdot \frac{(e^2/64)}{4\pi\epsilon_0} [\frac{3}{c} + \frac{3}{c\sqrt{2}} + \frac{1}{c\sqrt{3}}] = +\frac{e^2}{4\pi\epsilon_0} \frac{3 + 3/\sqrt{2} + 1/\sqrt{3}}{32R} = +\frac{e^2}{4\pi\epsilon_0} \frac{0,177}{R} \quad \text{(B.2)}$$

The sum of the various contributions is negative. From this electrostatic stand-point, the structure can be considered as stable. The sum is equal to:

$$E_s = -0,4\frac{e^2}{4\pi\epsilon_0 R} \quad \text{(B.3)}$$

c) The total Coulomb energy of a crystal of N cells includes, in addition, the electrostatic energy between the different cells. Each cell has zero electric charge and no electric dipole since the center of gravity of the negative charges coincides with that of the positive charges, at the center of the cell. Hence, each cell contains only electric multipoles of order at least equal to four. For such high order multipoles the interacttion energy decreases rapidly with the distance, and it will therefore be much smaller than NE_s.

2°) a) The chlorine ions are situated at the vertices of the cube and at the center of its faces. Teh sodium ions are at the center of the cell and at the center of its edges. In the structure, the minimum distance between chlorine ions equals one half of a face diagonal i.e. $c'/\sqrt{2} = 2R$. The cubic cell is a multiple cell (its volume equals four times that of a primitive cell) and a choice of primitive translations is, for instance:

$$\vec{a_1} = \frac{c'}{2}(\vec{i} + \vec{j}) \; ; \; \vec{a_2} = \frac{c'}{2}(\vec{j} + \vec{k}) \; ; \; \vec{a_3} = \frac{c'}{2}(\vec{i} + \vec{k}) \quad \text{(B.4)}$$

(Other choices are possible which all correspond to the same cell volume).

b) The most stable structure, which has the most negative energy, is the FCC structure. It is indeed the one which is experimentally observed for sodium chloride (see hereunder the detailed calculation for E_c).

Addendum

In the second structure (right of Fig. B.1) there are two contributions to the negative part of the Coulomb interaction: a) Sodium ions interacting with the chlorine ions at the vertices and b) interacting with the chlorine ions situated at the center of the faces. The first contribution involves, for each of the 8 chlorine ions an interaction with 13 sodium ions (12 having a charge $+1/4$ and one with charge $+1$).

$$8.\frac{-e^2/32}{4\pi\epsilon_0}[\frac{4}{c'\sqrt{3}/2} + \frac{3}{c'/2} + \frac{6}{c'\sqrt{5}/2} + \frac{3}{3c'/2}] = -\frac{4,49e^2}{4\pi\epsilon_0 c'} \quad \text{(B.5)}$$

The second contribution involves for each of the 6 chlorine ions an interaction with the 13 sodium ions:

$$6.\frac{-e^2/8}{4\pi\epsilon_0}[\frac{4}{c'/2} + \frac{4}{c'/2} + \frac{4}{c'\sqrt{3}/2} + \frac{4}{c'\sqrt{5}/2}] = -\frac{18,15e^2}{4\pi\epsilon_0 c'} \quad \text{(B.6)}$$

The sum of the Coulomb interactions is, finally $-22,64e^2/(4\pi\epsilon_0 c')$.

The Coulomb repulsion energy, determined by the same method is $(+17,30e^2/4\pi\epsilon_0 c')$. Indeed, the repulsion between sodium ions, between chlorine ions situated at the vertices, between chlorine ions situated at the center of the faces, and finally between the two types of chlorine ions are respectively (up to the factor $e^2/4\pi\epsilon_0 c'$):

$$+(\frac{12}{2})(\frac{1}{16})(\frac{4}{\sqrt{2}/2}+2+\frac{4}{\sqrt{3}/2}+\frac{1}{\sqrt{2}}+\frac{2.4}{\sqrt{2}/2}) = 8,604 \tag{B.7}$$

$$(\frac{8}{2})(\frac{1}{64})(3+\frac{3}{\sqrt{2}}+\frac{1}{\sqrt{3}}) = 0,356 \tag{B.8}$$

$$(\frac{6}{2})(\frac{1}{4})(\frac{4}{1/\sqrt{2}}+1) = 4,992 \tag{B.9}$$

$$8.\frac{1}{16}(\frac{3}{\sqrt{2}/2}+\frac{3}{\sqrt{3}/2}) = 3,346 \tag{B.10}$$

The sum of all the contributions is:

$$-5,34e^2/(4\pi\epsilon_0 c') = \frac{-1,89e^2}{4\pi\epsilon_0 R} \tag{B.11}$$

Exercise 2. *Crystal structure and mechanical properties of solids*

1°) a) BCC: $\vec{a_1} = \frac{c}{2}(\vec{i}+\vec{j}-\vec{k})$; $\vec{a_2} = \frac{c}{2}(-\vec{i}+\vec{j}+\vec{k})$; $\vec{a_3} = \frac{c}{2}(\vec{i}-\vec{j}+\vec{k})$.
Volume of the primitive cell $(c^3/2)$

As a reminder, FCC structure: $\vec{a_1} = \frac{c}{2}(\vec{i}+\vec{j})$; $\vec{a_2} = \frac{c}{2}(\vec{j}+\vec{k})$; $\vec{a_3} = \frac{c}{2}(\vec{i}+\vec{k})$

Other sets of primitive translations are possible. The choice made involves vectors which can be related by circular permutations and constituting a right handed reference frame. In an FCC structure one could choose, for instance $c/2(\vec{i}-\vec{j})$, $c/2(\vec{j}-\vec{k})$, $c/2(\vec{k}+\vec{i})$.

Note that there is an infinite choice of primitive translations. Each set generates all the nodes of the Bravais lattice. Indeed, if $(\vec{u},\vec{v},\vec{w})$ are a set of primitive translations, $(\vec{u}+n\vec{v}, \vec{v}, \vec{w}+m\vec{v})$ will be another set with the same unit cell volume.

b) The distances between atoms (equal to two atomic radii) are equal to the moduli of the primitive translations. Hence $r_{Fe} = c_{Fe}\sqrt{3}/4 = 1,24\text{Å}$ $r_{Cu} = c_{Cu}\sqrt{2}/4 = 1,28\text{Å}$. The volume of an atom is $(4\pi r^3/3)$. The ratio between the atomic volume and that of the cell is:

$$\frac{V_{Fe}}{V_{CC}} = \frac{\pi\sqrt{3}}{8} = 0,68 \qquad \frac{V_{Cu}}{V_{FCC}} = \frac{\pi\sqrt{2}}{6} = 0,74 \tag{B.12}$$

The most compact structure is that of copper. The coordinance of an atom is 12 (12 nearest neighbours) while in iron it is only 8. Nevertheless, copper is less rigid than iron. Compactness is thus not the only relevant element which determine the elastic stiffness or the hardness of a solid. As explained in chapter 8, the hardness (measured by the yield strength) is even generally smaller for compact structures.

c) The dense planes of the FCC structure are perpendicular to the cube main diagonals. In these planes the distance between atoms is equal to the modulus of the primitive translations. The glide directions are those of the primitive translations. For instance in the compact plane normal to the (111) direction these are: $\pm\vec{u}_1 = \pm(c/2)(\vec{i} - \vec{j}); \pm\vec{u}_2 = \pm(c/2)(\vec{j} - \vec{k}); \pm\vec{u}_3 = \pm(c/2) \ (\vec{i} - \vec{k})$.

In the BCC structure the dense planes are normal to the directions $(\vec{i} \pm \vec{j}); (\vec{j} \pm \vec{k}); (\vec{i} \pm \vec{k})$. They are defined by the diagonals (parallel to each other) of two opposite faces. Hence, there are 6 equivalent such compact planes. In each plane there are two glide directions parallel to the main diagonals of the cubic cell.

d) Among the twelve nearest neighbours, all situated at the distance $c_{cfc}\sqrt{2}/2 \tilde{~} 0,707 c_{cfc}$, 6 are located in the same compact plane (e.g. passing through the origin) and they form a regular hexagon generated by the six translations $\pm\vec{u}_1; \pm\vec{u}_2; \pm\vec{u}_3$, indicated in c). three are situated in one of the adjacent compact planes and the three others in the other adjacent compact plane. Indeed the three neighbouring atoms in the B plane are located at the positions $\vec{v}_1 = (c/2)(\vec{i} + \vec{j}), \vec{v}_2 = (c/2)(\vec{j} + \vec{k}); \vec{v}_3 = (c/2)(\vec{i} + \vec{k})$. The three neighbours of the other adjacent plane are located at $-\vec{v_1}, -\vec{v}_2, -\vec{v}_3$. A given atom and its three neighbours in the adjacent plane form a regular tetrahedron since $|\vec{v}_i| = |\vec{v}_i - \vec{v}_j| = c/\sqrt{2}$.

2°) a) The distance in the plane (triangular lattice) is:

$$|\overrightarrow{OO'}| = \mathbf{a} \Rightarrow \mathbf{c} = \mathbf{2a}\sqrt{\frac{2}{3}} \tag{B.13}$$

This relationship is approximately fulfilled in titanium (c/a=1, 59 instead of 1,63). There are two atoms of radius $\mathbf{a/2}$ in each primitive cell, whose volume is $(\vec{a_1}, \vec{a_2}, \vec{c}) = \mathbf{a}^2 c\sqrt{3}/2$. The ratio between the volume of the atoms and the volume of the crystal is thus:

$$\frac{2V_{Ti}}{V_{hex}} = \frac{\pi\sqrt{2}}{6} = 0,74 \tag{B.14}$$

Additional remark: The identical compactness of the FCC and hexagonal structures is due to the fact that both structures are a stacking of identical atomic planes, only differing in their sequence: ABCABC... for the FCCc structure and ABABAB... for the hexagonal one (cf. chapter 2). Alternately, one can note that the lateral shift between two adjacent planes is the same in the two structures, respectively: $(1/3)(\overrightarrow{a}_1 + \overrightarrow{a}_2) = \overrightarrow{OO'} - \overrightarrow{c}/2$ and $\overrightarrow{g} = -(1/3)(\overrightarrow{u}_1 + \overrightarrow{u}_3)$. These two vectors are equal, up to a change of sign.

3°) The primitive translations are the same as in copper. The distance between neighbouring atoms is $|(\mathbf{c}/4)(\overrightarrow{i} + \overrightarrow{j} + \overrightarrow{k})| = |\overrightarrow{a_i} - (\mathbf{c}/4)(\overrightarrow{i} + \overrightarrow{j} + \overrightarrow{k})| = \mathbf{c}\sqrt{3}/4$. Each atom has 4 neighbours forming a tetrahedron. The atomic radius is $\mathbf{c}\sqrt{3}/8 = 0,77$Å. The filling factor is then:

$$\frac{2V_C}{V_{CFC}} = \frac{\pi\sqrt{3}}{16} = 0,34 \qquad (B.15)$$

Diamond is the stiffest and hardest of the considered solids. However it is the less compact. The major factor in its mechanical properties consists in its covalent bonds which are have little tolerance towards variations of distances and of angles (cf. chapter 7, Fig. 7.6).

Exercise 3. *Quenching and annealing of vacancies in gold*

1°) The lengthening of the wire is due to the presence of vacancies. To each vacancy one associates the volume $(2\omega + \Omega)$: $\delta\Omega$ is the positive or negative increment of volume of the vacancy with respect to the volume ω of the missing atom. One must also take into account the volume of the atom which migrates to the surface and, nevertheless remains in the wire.

Vacancies exist at thermal equilibrium, because their positive formation energy is finite. Their number increases when the temperature of the sample is raised. This is the reason of the observed lengthening of the wire. If the wire is not quenched, the number of vacancies would regress, on cooling, to its initial value and the length of the wire would also be restored. The quenching is assumed to "freeze" the number of vacancies to the equilibrium value at the highest temperature reached. At room temperature, the migration of atoms from the surface to the vacancy sites is extremely slow and will not restore in practice the equilibrium concentration of vacancies. On the other hand, quenching has the advantage of eliminating another cause of lengthening of the wire, i.e. its thermal expansion.

2°) a) Let l_0 be the initial length of the wire, l its length after quenching, and $\delta = (l - l_0)$:

$$\left(\frac{l}{2}\right)^2 = \left(\frac{l_0}{2}\right)^2 + h^2 \quad \rightarrow \quad \frac{\delta l}{l_0} = \frac{2h^2}{l_0^2} = \frac{1}{3}\frac{\delta V}{V_0} = \frac{n_{lacunes} \cdot (\omega + \delta\Omega)}{3N_{atomes} \cdot \omega} = \frac{\mathbf{c}(\omega + \delta\Omega)}{3\omega} \qquad (B.16)$$

then:

$$h = l_0 \sqrt{\frac{(\omega + \delta\Omega)}{6\omega}} \cdot \sqrt{c} \qquad (B.17)$$

in which ω is the volume of an atom, and $\delta\Omega$ is the difference of volume between a vacancy and the substituted atom.

b) In Fig. 2 the logarithm of h varies linearly as a function of $(1/T)$. Referring to chapter 4, Eq. (4.5), one expects that:

$$c = \frac{n_{lac}}{N} = \exp(-\frac{u_F}{k_B T}) \quad \rightarrow Ln(h) = B - \frac{u_F}{2k_B \cdot T} \qquad (B.18)$$

Where k_B is the Boltzmann constant. The negative slope of $Ln(h)$, in the figure provides the following values:

$$Ln(\frac{h_1}{h_2}) = Ln(\frac{550}{9,5}) = 4,06 = (\frac{u_F}{2.10^3 k_B})(\frac{10^3}{T_2} - \frac{10^3}{T_1}) \approx 4,06 u_F (\text{eV}) \qquad (B.19)$$

From which the value of u_F can be drawn:

$$u_F \approx 1eV \qquad (B.20)$$

c) At the melting temperature $c = \exp(-u_F/k_B.1337)$ $(k_B.1337(\text{eV}) = 1,38.10^{-23}.1337/1,602.10^{-19}) = 0.115$ eV . Hence:

$$c = \exp(-1/0.115) \approx 1,67.10^{-4} \qquad (B.21)$$

Comment: This result corresponds approximately, in each direction to one atom missing out of ten. This is sufficient to destroy the cohesion of the crystal: One of the theories of melting assigns this phenomenon to the increase of the number of vacancies beyond a certain concentration.

3°) a) This variation can be deduced from Eq. (B.17). For $(10^3/T) = 0,9$ and $h = 300\mu$. $c = \exp(-u_F/k_B T) \approx 3.10^{-5}$, thus:

$$\frac{\delta\Omega + \omega}{\omega} = \frac{6.h^2}{c.l_0^2} = \frac{6.9.10^{-4}}{3.10^{-5}.4.10^2} = 0,45 \qquad (B.22)$$

and $\delta\Omega = -0,55\omega$. A vacancy induces, in the solid, a local contraction of the volume at the site previously occupied by an atom. This result is consistent for solids made of monovalent atoms. (Although gold has a chemical valency equal to 3, its "metallic valency" is equal to 1). The "radius" b of the vacancy is 0,77a. (a is the radius of the atom).

b) The deformation field and the elastic energy are provided by formulas (3.13) and (3.14) of chapter 3. In particular, $W = 6\mu\omega(b-a)^2/b^2$, with $\omega = 6,4.10^{-30}m^3$, and $(b-a)^2/b^2 \approx 5.10^{-2}$. Hence $W = 12.10^{10}.6,4.10^{-30}.5.10^{-2}$ Joules $\approx 0,25eV \approx \frac{1}{4}u_F$. Note that this value which derives from the "elastic model" of the vacancy-energy does not match well the actual value of the formation energy of the vacancy. This means that the main part of the formation energy is rather related to a change of the electronic energy of the bonds between atoms in the neighbourhood of the vacancy.

4°) a) The vacancies diffuse slowly towards the surface and disappear. The concentration then recovers its equilibrium value at $40°C$.

b) The jump rate of the vacancies is (chapter 4, Eq. (4.15)):

$$f = \nu z e^{-\beta E_M} \approx 10^{13} z e^{-\beta E_M} \tag{B.23}$$

where E_M is the migration energy of the vacancies and z the number of nearest neighbours of an atomic site.

This rate is related to the diffusion coefficient of the vacancies. In an FCC structure, one has (cf. chapter 4, Eq. (4.24)), with $\delta^2 = a^2/2, z = 12, \gamma = 1/6$): $D_l = \nu a^2 \exp(-\beta E_M) = 16.10^{-7} \exp(-\beta E_M)$.

5°) a) The diffusion equation is:

$$\frac{\partial c}{\partial t} = D_l.\Delta c \tag{B.24}$$

Which reduces, in one-dimension to:

$$\frac{\partial c(z,t)}{\partial t} = D_l.\frac{\partial^2 c}{\partial z^2} \tag{B.25}$$

Limiting conditions: The vacancies and atoms recombine on the lateral surfaces of the wire, hence the concentration of vacancies is zero for any value of t. $c(t) = 0$ for z=0 and z=d (the two ends of the diameter of the wire). For $t = 0$, the concentration in the entire wire (except on the lateral surfaces) has the value reached just after the quenching: $c = c_0$; For $t = \infty$, the concentration must tend towards ≈ 0 (the equilibrium concentration at $40°C$ can be neglected when compared to c_0).

b) It is easy to check that the above diffusion equation has, as a solution, the suggested expression. It yields: $\kappa = \pi/d \approx 3.10^4 m^{-1}$ and $A = c_0$. At any time t the spatial distribution of the vacancies has a "sine-arch" shape. For increasing times, the height of this arch lowers proportionally to $\exp(-\kappa^2 D_l t)$.

c) The experimental curve reaches a value $\approx c_0/e$ after approximately 200 hours=$7,25.10^5$s. Hence, at $40°C$:

$$D_l = \frac{1}{\kappa^2 t} = \frac{1}{9.7,25.10^{13}} \approx 1,5.10^{-15} m^2.s^{-1} = \nu a^2 e^{-\beta E_M} \qquad (B.26)$$

And, consequently, taking $\nu = 10^{13} s^{-1}$ and $a = 4.10^{-10} m$ ::

$$E_M \approx 0,56 \text{eV} \qquad (B.27)$$

6°) The slowness of the cooling process allows to reach thermal equilibrium at each temperature. The concentration of vacancies thus decreases progressively to its almost zero value at room temperature.

7°) Referring to chapter 3 (Eq. (1.28)) the activation energy relative to the migration of atoms is $(E_M + u_F)$=1,55eV. Thus, $D_A = c_l D_l = \nu a^2 \exp(-\beta[E_F + E_M]) \approx 5.10^{-14} m^2.s^{-1}$. This value can be compared to the diffusion coefficient of vacancies at the same temperature: $\approx 2,5.10^{-11} m^2.s^{-1}$.

Exercise 4. *Doping by diffusion of a semiconductor*

1°) Silicon has a structure of the diamond type.

$$D_{auto} \propto \exp\left(-\beta[E_M + u_F]\right) \qquad (B.28)$$

Thus the logarithm of D is linear as a function of $(1/T)$. From Fig. A3 (left) one draws:

$$(E_M + E_F) = 10^4.k_B Ln(\frac{D_2}{D_1})/(\frac{10^4}{T_1} - \frac{10^4}{T_2}) = 4,3 \text{eV} \qquad (B.29)$$

2°) In the simplest situation (substitution of silicon by boron) the same relationship holds but with a different migration energy. From Fig. A3 (right) one deduces:

$$(E_M^B + E_F) = 3,3 \text{eV} \to E_M - E_M^B = 1 \text{eV} \qquad (B.30)$$

Note that the migration energy of boron is smaller, consistently with its smaller size (0,8Å instead 1,1Å). However if the values of D at the same temperature are compared it appears that, at $1200°C$, for instance, one has :

$$\frac{D_B}{D_{Si}} = 10^2 << (\nu_B/\nu_{Si}) \exp(\beta[E_M - E_M^B]) \approx 5.10^3 \qquad (B.31)$$

This result probably comes from the fact that the boron atoms interact repulsively with the silicon vacancies (in this case the diffusion coefficient will not be proportionnal to the vacancies concentration).

3°)

$$c(x,t) = c_s \left[1 - \frac{2}{\sqrt{\pi}} \int_0^{\frac{x}{\sqrt{4Dt}}} e^{-\lambda^2} . d\lambda \right] \tag{B.32}$$

Let $f(u) = \exp(-u^2)$ with $u = x/\sqrt{4Dt}$. Then $\partial c/\partial t = Af(u)(-u/2t)$ and $\partial^2 c/\partial x^2 = Af'(u)(u/x)^2$. Since $f'(u) = -2uf(u)$ one can check that:

$$\{-Af(u)\frac{u}{2t} = -Af(u)\frac{x}{2t\sqrt{4Dt}}\} = \{-2DAuf(u)(\frac{u}{x})^2 = -Af(u)\frac{2Dx}{4Dt\sqrt{4Dt}}\} \tag{B.33}$$

On the other hand, the limit conditions are also satisfied (x=0 c=c_s, $t = 0$ c=0 except for x=0).

$$\frac{2}{\sqrt{\pi}} \int_0^X \exp(-u^2) du \tag{B.34}$$

Assume that the penetration at a depth 2μ corresponds to the value $0{,}5c_s$. This implies a value of the error function equal to 0.5. This function has the value 0.5 for X ˜0.5. Hence:

$$\frac{x}{\sqrt{4Dt}} = 0.5 \rightarrow t = \frac{x^2}{D(T)} \tag{B.35}$$

For T=1000°C ($D = 10^{-14}$cm^{2}s^{-1}), t=4.10^6s = 45 days. For T=1200°C, t is equal to about ten hours.

Exercise 5. *Energy of a vacancy and heat capacity of a solid*

1°) Each vacancy is associated to an increase of energy u_F. The contribution ΔU of the vacancies is thus: $\Delta U = n.u_F$. The concentration of vacancies at a temperature T is $c = (n/N) = \exp(-u_F/k_B T)$ in which N is the number of atomic sites in the crystal which are normally occupied by a gold atom, and which are likely to be substituted by a vacancy. (Since $n \ll N$, it is meaningless to include in N the number of surface sites towards which the atoms migrate when substituted by a vacancy.) Hence:

$$\Delta U = N.u_F.e^{-(u_F/k_B T)} \tag{B.36}$$

2°) The excess of heat capacity of the sample, due to the vacancies, is $\Delta C = d(\Delta U)/dT, N$ thus:

$$\Delta C = \frac{d(n.u_F)}{dT} = \frac{Nu_F^2}{k_B T^2} \exp(-u_F/k_B T) \tag{B.37}$$

and:

$$Ln(T^2 \Delta C) = Ln(\frac{N u_F^2}{k_B}) - \frac{u_F}{k_B T} \tag{B.38}$$

In agreement with Fig. A.4 (b) it is linear as a function of $(1/T)$. From the slope in Fig. A.4 (b) one can deduce the value of u_F :

$$\frac{u_F}{10^4 k_B}(\frac{10^4}{T_1} - \frac{10^4}{T_2}) = Ln(T_2^2 \Delta C_2) - Ln(T_1^2 \Delta C_1) \tag{B.39}$$

Thus, using the temperature interval indicated in Fig. A.4 (b):

$$\frac{u_F}{0,86}(10 - 8) \approx (12, 2 - 9, 8) \rightarrow u_F \approx 1.03 \text{eV} \tag{B.40}$$

3°) The crystal contains one gold atom per primitive unit cell (whose volume is one fourth of that of the conventional cubic cell). The volume of the sample is thus equal to the volume of the primitive cell times the number N of gold atoms (N is the number of sites considered hereabove): $\Omega = 16.N$ Å^3 = $16N.10^{-24}$cm^3. N is determined by the coordinates of any of the points on the straight line representing graphically $Ln(T^2 \Delta C)$.For instance, chosing the point corresponding to T_2:

$$Ln(\frac{k_B T_2^2 \Delta C}{N u_F^2}) = -\frac{10^4 . u_F}{10^4 k_B T_2} \rightarrow Ln(\frac{u_F}{10^4 k_B}.\frac{10^4}{T_2}.\frac{N.u_F}{T_2 \Delta C}) = 9.6 \tag{B.41}$$

or:

$$N = \frac{T_2^2 \Delta C}{u_F . T_2} \frac{e^{9,6}}{9,6} \approx 1.5.10^{24} \rightarrow \Omega = 16.10^{-24} N \approx 24 \text{cm}^3 \tag{B.42}$$

Exercise 6. *Straight dislocation and plastic deformation*

1°) If Fig. A.5 in the text is ignored, several answers are possible (cf. chapter 5). Actually Fig. A.5 shows that the dislocation has been formed through a slip of the upper left part of the crystal. This slip is uniform except in the vicinity of the dislocation line where it is blocked. This line is therefore a line of discontinuity of the displacement. The amplitude of the relative slip is equal to b, i.e. the modulus of the Burgers vector. The right upper part of the crystal and the lower part of the crystal are unmoved.

Remark: After suppression of the stresses which induce the formation of the dislocation, the solid evolves towards a state in which the strain field becomes symmetric.

2°) A step of width b occurs on the surface through which the dislocation is evacuated (Fig. B.3).

3°) Considering the small value of the width of the step (a few interatomic spacings), the preceding figure can be identified with the sheared parallelipiped represented hereunder (Fig. B.4 by dotted lines.)

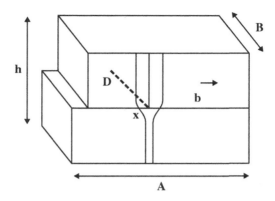

Figure B.2

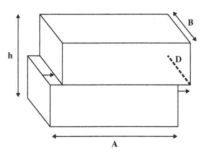

Figure B.3

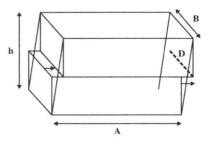

Figure B.4

The induced deformation is a shear corresponding to an angle between A and h. Hence:

$$\epsilon = \frac{1}{2}\sin\theta = \frac{b}{2h} \tag{B.43}$$

4°) It is reasonable to assume that the deformation is proportional to the fraction of the surface which has slipped. Its value is therefore:

$$\epsilon = \frac{x}{A}\cdot\frac{b}{2h} \tag{B.44}$$

5°) In the case of a screw dislocation, there is a similar ambiguity concerning the nature of the relative slip. If one choses the case represented in Fig. A6, it is the upper part of the solid which has slipped by an amplitude equal to b.

The gliding of the dislocation throughout the entire width of the sample induces a slip of the upper part of the crystal which is identical to that obtained in the case of the edge dislocation (note that the Burgers vectors of the two dislocations are identical but that the dislocation lines are normal to eachother). A shear of same characteristics and magnitude is thus determined. A partial slip determines a shear proportional to (x/B).

Exercise 7. *Rotation of a straight dislocation having a fixed point*

a) CD and DE are dislocations (cf. Fig. A.7) because they are are both lines separating a part of a solid which has slipped (the part above ABCD) and a part which is unmoved (behind CD): they are lines of discontinuity of the displacement. They are edge dislocations since the displacement b of the slip is normal to the direction of the lines. The compressed region of the solid is above CD and at the right of DE. As a consequence, the portion of lattice plane whose insertion in a crystal can be associated to this L-shaped is the quarter of a plane CDE (cf. Fig. B.5)

b) When CD is rotated by 45° (cf. Fig. A.7 (right)), it becomes a mixed dislocation line (partly edge- and partly screw-). The Burgers vector has components parallel and perpendicular to the dislocation line. If the value of the rotation angle is increased, CD will eventually be parallel to the slip direction, and the dislocation becomes of pure screw-type. orienting the dislocation from D towards C, a Burgers closed loop,, seen from C as right-handed, defines a Burgers vector parallel to $\overrightarrow{DC}$ (As confirmed by Fig. B.6 hereunder in which matter is followed by continuity by progressing from D to C). Thus, the dislocation is right handed. On the other hand, the convention defined in chapter 5 for edge-dislocations, defines a right handed frame of reference (cf. right of fig.B.6) formed a) by the direction $\overrightarrow{b}$ of the Burgers vector (parallel to $\overrightarrow{AD}$), b) by the vector pointing towards the additional half plane mentioned above, and c) by the oriented dislocation. The latter direction is therefore oriented from E to D, in continuity with the sense taken for the screw dislocation DC.

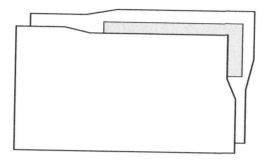

Figure B.5

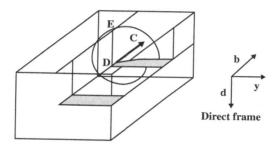

Figure B.6

c) AD (Fig. A.7) is not a dislocation as it is not a line of discontinuity. Indeed, the same translation defines the gliding of the two planes EDAF and ABCD. AD would be a dislocation if ABCD slipped while EDAF did not slip (which precisely occurs beyond D).

Exercise 8. *Dislocations and curvature of a rod*

In the ideal crystal structure which is used to determine the closure defect one has A'B'=C'D'. The closure defect δ is therefore equal to (CD-AB), or, as $CD = (R + BC)\theta$ and $AB = R\theta$:

$$\delta = BC.\theta = \frac{BC.AB}{R} \qquad (B.45)$$

Each dislocation induces a closure defect equal to b. The number of dislocations is thus δ/b and their density:

$$\rho = \frac{\delta}{b.AB.BC} = \frac{1}{bR} \tag{B.46}$$

The numerical application leads to $\rho \approx 4.10^6 cm^{-2}$.

Exercise 9. *Intersection of dislocations in an FCC structure*

Orienting the lines towards their intersection one can write (cf. chapter 5, section 5.3.3):

$$\sum \vec{b}_i = 0 \tag{B.47}$$

In order to satisfy this equation a minimum number of vectors, having equal modulus, must be found among primitive translations. In the FCC cubic structure one needs at least 3 dislocations with Burgers vectors forming equilateral triangles. These dislocations will be edge-, screw- or mixed- dislocations.

Exercise 10. *Dislocations in a hexagonal structure*

1°) $\vec{d}_1$ et $\vec{d}_2$ are not primitive translations because if the origin of these vectors is set on an atom of the structure, their extremity does not always end on another atom. Within the A hexagonal planes, the primitive translations are, for instance:$(2\vec{d}_1 + \vec{d}_2) = (3a/2)\vec{i} + (a\sqrt{3}/2)\vec{j}$ et $(\vec{d}_1 + 2\vec{d}_2) = a\sqrt{3}\vec{j}$. In these planes, the basis is composed of two atoms related to each other by a vector $(\vec{d}_1 + \vec{d}_2) = (a/2)\vec{i} + (a\sqrt{3}/2)\vec{j}$.The third primitive translation is equal to $\vec{c}$. The three-dimensional basis comprises, in addition, the two atoms of a B plane related to the basis of the A plane by the translation $(\vec{c}/2 + \vec{d}_1)$.The cell volume is $V = 3a^2 c\sqrt{3}/2$.

2°) Applying the definition of the reciprocal lattice leads to :

$$\vec{b}_1^* = \frac{4\pi}{a}\vec{i}; \vec{b}_2^* = \frac{2\pi}{a\sqrt{3}}(-\frac{\vec{i}}{\sqrt{3}} + \vec{j}); \vec{c}^* = \frac{2\pi}{c}\vec{k} \tag{B.48}$$

3°) The Burgers vectors are the primitive translations in these planes, i.e. the $\vec{b}_i$ and the similar vectors. The vector normal to the A planes is parallel to $\vec{c}$ and $\vec{c}^*$. Its is therefore perpendicular to the Burgers vectors of the dislocations, and hence the dislocations will not be visualized by the experiment (cf. chapter 5, section 5.5).

Exercise 11. *Observation of dislocations by electron microscopy*

1°) a) The families of parallel lattice planes whose spacing $d_1 = c\sqrt{3}/3$ is largest are also the ones composed of the most dense planes. These are the planes perpendicular to the diagonals of the cubic cell. Among the lattice planes parallel to $\vec{c}_3$,the ones with the largest spacing are parallel to the edges of the cubic cell and passing by the vertices and the centers of the faces (distance $d = c/2$). A slightly smaller distance $(c\sqrt{2}/4)$ separates the planes parallel to the face diagonals.

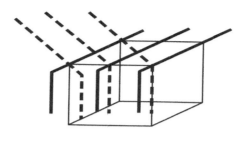

Figure B.7

b) The reciprocal lattice is defined by its relations to the primitive translations:

$$\vec{a}_i^* = \frac{2\pi \vec{a}_j \wedge \vec{a}_k}{(\vec{a}_1, \vec{a}_2, \vec{a}_3)} \tag{B.49}$$

Taking into account the equality $(\vec{a}_1, \vec{a}_2, \vec{a}_3) = (c^3/4)$:

$$\vec{a}_1^* = 2\pi \frac{(c^2/4)(-\vec{i} + \vec{j} + \vec{k})}{(c^3/4)} = \frac{2\pi}{c}(-\vec{i} + \vec{j} + \vec{k}) \tag{B.50}$$

and, by circular permutations: $\vec{a}_2^* = (2\pi/c)(\vec{i} - \vec{j} + \vec{k})$, $\vec{a}_3^* = (2\pi/c)(\vec{i} + \vec{j} - \vec{k})$. The lattice generated by these vectors is of the cubic centerd type (BCC) with a conventional cubic cell having an edge equal to $(4\pi/c)$.

c) Each family of equally spaced parallel lattice planes contains all the nodes of the Bravais lattice (cf. chapter 2). One can therefore obtain a set of primitive translations of a crystal by chosing two vectors $(\vec{b}_1, \vec{b}_2)$ joining neighbouring nodes of a given lattice plane and chosing as third vector $\vec{b}_3$ joining any two nodes belonging to adjacent planes of the same family. Clearly these three vectors will generate the entire set of Bravais lattice nodes. On the other hand, if $\vec{n}$ is the normal common to the former lattice planes one can write $\vec{b}_3.\vec{n} = d$ (interplane spacing). The third vector of the reciprocal lattice defined by the $\vec{b}_i$ is:

$$\vec{b}_3^* = 2\pi \frac{\vec{b}_1 \wedge \vec{b}_2}{V} \tag{B.51}$$

It is perpendicular to the lattice planes since it can be checked that $\vec{b}_3^* = a\vec{n}$. Besides, $\vec{b}_3^*.\vec{b}_3 = 2\pi = a\vec{b}_3.\vec{n} = ad$, Hence:

$$\vec{b}_3^* = \frac{2\pi}{d}.\vec{n} \tag{B.52}$$

$\vec{b}_3^*$ is the $\vec{g}$ vector required.

2°) a) The Bragg condition implies that a reflection can occur on a family of lattice planes. It expresses the fact that the incident beam is at the "Bragg" incidence satisfying $2d \sin \theta = h\lambda$. Equivalently, the vector $(\vec{q}' - \vec{q})$ must be equal to a $\vec{G}$ vector belonging to the reciprocal lattice.

b) The main reason explaining that several beams can be reflected simultaneously is the small value of the wavelength of the electrons, as compared to the modulus of the primitive translations ($\lambda/c \approx 10^{-2}$). The condition $(\vec{q}' - \vec{q}_1) = \vec{G}$ then implies that $\vec{G}$ is almost perpendicular to the wavevectors of the incident and reflected beams. One can adjust the orientation of the crystal in such a manner as to position a plane of the reciprocal lattice perpendicular to $\vec{q}_1$. The condition $(\vec{q}' - \vec{q}_1) = \vec{G}$ can then be fulfilled simultaneously for several $\vec{G}$ vectors of this reciprocal plane, (Fig. B.8).

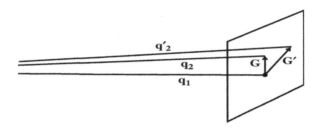

Figure B.8

3°) a) For reasons indicated in 2°) b), the focal plane of the microscope, normal to $\vec{q}$, is parallel to the reciprocal plane containing the vectors $\vec{G}$ corresponding to the different reflected directions. Each vector $\vec{G}$ is normal to a family of parallel lattice planes. On the other hand, as shown by Fig. B.9, the distance between the origin and the node, in the focal plane, corresponding to a given reflection, is (for a small θ angle), $fh\lambda/d$, where f is the focal length, and h is the integer number appearing in the Bragg equation $2d \sin \theta = h\lambda$. This distance is proportional to G. The focal plane of the microscope thus contains an image of the considered reciprocal plane (with a magnification equal to $f\lambda$).

b) The observed spots reproduce the configuration of nodes of the reciprocal plane normal to $\vec{c}_3^*$. The shortest vectors $\vec{G}_i$ of the reciprocal lattice which are perpendicular to $\vec{c}_3^*$, are combinations of $(4\pi/c) \vec{i}$ and of $(4\pi/c) \vec{j}$. The coordinates of the observed spots are thus of the form $2h\vec{c}_1^* + 2k\vec{c}_2^* + 0.\vec{c}_3^*$. The central spot has indices $(0,0,0)$ while the neighbouring spots have indices $(\pm 2,0,0)$, $(0,\pm 2,0),(\pm 2,\pm 2,0)$.

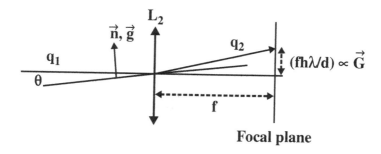

Figure B.9

4°) a) A dislocation will not be observable if $\vec{b}.\vec{G} = 0$, because, in this case, $\vec{G}$ is perpendicular to the planes involved in the diffraction. These planes are parallel to $\vec{b}$ and are therefore almost undeformed by the dislocation (cf. chapter 5, section 5.5)

b) See the table hereunder (V=visible, I=invisible).

$\dfrac{\vec{g}}{2\vec{b}}$	g_1	g_2	g_3	g_4	Identification of b	type
(110)	V	V	V	I		
$(1\bar{1}0)$	V	I	V	V		
(101)	V	V	I	V		
$(1\bar{1}0)$	V	V	I	V		
(011)	I	V	V	V		
$(01\bar{1})$	I	V	V	V		
a	V	I	V	V	$(1\bar{1}0)$	Almost edge
b	V	V	V	I	(110)	mixed
c	V	V	I	V	(101) ou $(10\bar{1})$	mixed
d	I	V	V	V	(011) ou $(0\bar{1}1)$	Almost edge
e	V	V	I	V	(101) ou $(10\bar{1})$	˜edge

Exercise 12. *Forces acting on a dislocation loop*

1°) a) The portions of the loop normal to b are of edge-type and the portions parallel to b are of screw-type. Considering the convention defined in chapter 5, the additional half plane is in the direction $\vec{s}$ such as $(\vec{b}, \vec{s}, \vec{d})$ be a direct-oriented frame, in which $\vec{d}$ is the local vector defining the sense of the dislocation loop. $\overrightarrow{s}$ is therefore oriented upwards for the edge dislocation located on the right, and downwards for the edge-dislocation located on the left. The screw dislocation situated in the front of the figure is left-handed ($\overrightarrow{b}$ is oriented opposite to $\overrightarrow{d}$), while the one on the back is right-handed.

b) The applied stress tends to reduce the spacing between the half-planes situated above the edge-dislocation situated at the right. Consequently, it tends to displace it to the right. Likewise, it tends to reduce the spacing between the half-planes situated below the edge dislocation situated at the left, and tends to displace it to the left. Another, equivalent, argument is to note that σ has the same sign as the stress which induces the slip $\overrightarrow{b}$. It will therefore tend to increase the glide surface. This argument holds also for the screw dislocations.

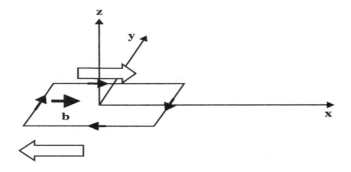

Figure B.10

Apply the Peach-Koehler relation to each section of the loop:

$$\overrightarrow{b}\overline{\overline{\sigma}} = (b,0,0) \begin{bmatrix} 0 & 0 & \sigma \\ 0 & 0 & 0 \\ \sigma & 0 & 0 \end{bmatrix} = \begin{pmatrix} 0 \\ 0 \\ b\sigma \end{pmatrix} \tag{B.53}$$

Consider, for instance, the edge-segment at the right and the screw-segment in front. The Peach-Koehler formula takes the form:

$$\overrightarrow{f}_E = \overrightarrow{b}\overline{\overline{\sigma}} \wedge \overrightarrow{dL} = \begin{pmatrix} 0 \\ 0 \\ b\sigma \end{pmatrix} \wedge \begin{pmatrix} 0 \\ -dL \\ 0 \end{pmatrix} = \begin{pmatrix} b\sigma dL \\ 0 \\ 0 \end{pmatrix} \tag{B.54}$$

$$\overrightarrow{f}_S = \overrightarrow{b}\overline{\overline{\sigma}} \wedge \overrightarrow{dL} = \begin{pmatrix} 0 \\ 0 \\ b\sigma \end{pmatrix} \wedge \begin{pmatrix} -dL \\ 0 \\ 0 \end{pmatrix} = \begin{pmatrix} 0 \\ -b\sigma dL \\ 0 \end{pmatrix} \tag{B.55}$$

The values being opposite for the two remaining dislocations. Hence, on each portion of the loop, the forces are oriented outwards, with respect to the surface of the loop. They will tend to increase its surface, in agreement with the preceding qualitative arguments.

Exercise 13. *Interaction between edge dislocations having parallel glide planes*

1°) a) The non-zero strain-components are:

$$\epsilon_{11} = -\frac{b\sin\theta(2 - 2v + \cos 2\theta)}{4\pi(1 - v)r}; \epsilon_{22} = \frac{b\sin\theta(2v + \cos 2\theta)}{4\pi(1 - v)r}; \epsilon_{12} = \frac{b\cos\theta\cos 2\theta}{4\pi(1 - v)r} \tag{B.56}$$

The stresses are determined by $\sigma_{ij} = 2\mu\epsilon_{ij} + \lambda\delta_{ij}.\delta$ or,

$$\sigma_{11} = -\frac{\mu b\sin\theta(2 + \cos 2\theta)}{2\pi(1 - v)r}; \sigma_{22} = \frac{\mu b\sin\theta\cos 2\theta}{2\pi(1 - v)r}; \tag{B.57}$$

$$\sigma_{33} = \frac{-2\mu bv\sin\theta}{2\pi(1 - v)r}; \sigma_{12} = \frac{\mu b\cos\theta\cos 2\theta}{2\pi(1 - v)r} \tag{B.58}$$

and:

$$\sigma_{rr} = \sigma_{\theta\theta} = -\frac{\mu b\sin\theta}{2\pi(1 - v)r}; \tag{B.59}$$

$$\sigma_{r\theta} = \frac{\mu b\cos\theta}{2\pi(1 - v)r}; \sigma_{zz} = -\frac{2v\mu b\sin\theta}{2\pi(1 - v)r} \tag{B.60}$$

b) One can write:

$$\delta = \sum \epsilon_{ii} = -\frac{\mu}{\lambda + 2\mu}\frac{b\sin\theta}{\pi r} \tag{B.61}$$

The compressed region ($\delta < 0$) corresponds to $0 < \theta < \pi$, the expanded one ($\delta > 0$) to $\pi < \theta < 2\pi$. The additional half-plane is in the compressed region. The dislocation being oriented towards the positive end of the z-axis, the standard convention defined in chapter 5 determines a Burgers vector oriented towards the positive end of the x-axis.

2°) Using the Peach-Koehler relation, one finds that the force per unit length of d_1 is , putting $D = \frac{\mu b}{2\pi(1-v)R}$:

$$\overrightarrow{b}\overline{\overline{\sigma}}\wedge\frac{\overrightarrow{z}}{z} = (b, 0, 0)\begin{pmatrix} \sigma_{11} & \sigma_{12} & 0 \\ \sigma_{12} & \sigma_{22} & 0 \\ 0 & 0 & \sigma_{33} \end{pmatrix}\wedge\begin{pmatrix} 0 \\ 0 \\ 1 \end{pmatrix} = bD\begin{bmatrix} \cos\theta\cos 2\theta \\ \sin\theta(2 + \cos 2\theta) \\ 0 \end{bmatrix} \tag{B.62}$$

a) The y component is positive in the upper half-plane and negative in the lower one. The x-component changes sign when going from one eigth angular

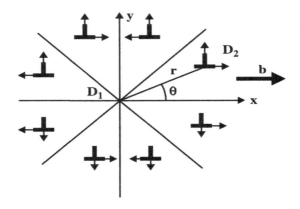

Figure B.11

fraction of the plane to the adjacent one. Whatever the value of θ the force is repulsive but its orientation changes as a function of the value of θ.

b) The glide plane is (xz). The dislocation d_1 is in equilibrium when the force is perpendicular to x. This is the case, for D_2, in the directions $\pm y$ and in the directions of the two bissectors. The positions $\pm y$ are stable, since a small deviation with respect to this direction tends to regress through the action of the tangential force. Indeed, the x-component of the force is oriented towards the y-axis. By contrast the positions on the bissectors are unstable: A small deviation will tend to be amplified. If the position is closer to the x-axis, it will tend to repell it to infinity. If it is closer to the y-axis, it will tend to bring it on the $\pm y$ axis.

c) The forces have a sense opposite to the one found above. If the displacement of D_2 is free, there is attraction to D_1 whatever the initial position. If only the displacement in the glide-plane is possible, the bissectors will be the stable positions. If D_2 is on the x-axis, it will move and annihilate on coinciding with D_1.

$3°$) The glide is the motion parallel to the glide plane (xz), while the climb is the motion perpendicular to the glide plane (yz). The results of $2°$) are used to evaluate the forces exerted by the two dislocations surrounding D_1. When located at the origin of coordinates this dislocation is submitted to opposite forces (equal in modulus) exerted by D_2 and D_{-2}. If the dislocation leaves the origin along the y-axis (thus performing a climb), The sum of the radial forces directed along this axis, tends to bring it back to the origin. If a small displacement occurs along x (this motion being a glide), the sum of the radial and tangential forces tends to bring the dislocation back to the origin (the situation is that encountered in 2°). However, if the displacement is large in this direction (i.e. larger than the half distance between the fixed dislocations), the dislocation D_1 tends to escape to infinity.

4°) a) Far from the surface, each dislocation is surrounded by $N/2$ pairs of dislocations whose effect is to keep it in equilibrium (this is a *cooperative* effect of stabilization).

b) Each dislocation is equivalent to the insertion of one additional half-plane. On either sides of the y-axis there is a regular succession of shifts each having the amplitude b, and the shifts being spaced by d. These shifts induce a difference of orientation whose slope is b/d.

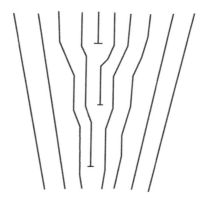

Figure B.12

c) The energy of an edge-dislocation per unit length (in the z-direction) is $(\mu b^2/4\pi(1-v)).\log{(d/b)}$. There are $(1/d)$ dislocations par unit length in the y-direction. One has therefore per unit area:

$$W = \frac{\mu b^2}{4\pi(1-v)d} \log(\frac{d}{b})$$

(B.63)

Thus $\theta = b/d = 1,6.10^{-2} \Rightarrow (d/b) = 60 \Rightarrow w = 10^{-1}J/m^2 = 0.1eV//cell.$

Exercise 14. *Interaction between screw dislocations*

The first dislocation is along the z-axis, with $\overrightarrow{b}$ parallel to $\overrightarrow{oz}$. The second one has a direction parallel to $\overrightarrow{b}$ and a Burgers vector $\overrightarrow{b'} = (0, b'/\sqrt{2}, b'/\sqrt{2})$. With the suggested notations, one has $\overrightarrow{PM} = (a, u/\sqrt{2}, u/\sqrt{2})$ with u varying in the range $(-\infty, +\infty)$.

The stress field induced by the screw-dislocation D at point M of the D'dislocation (Eq. (6.7) of chapter 6) has two components: $\sigma_{xz} = -\mu b y_M / 2\pi r^2$ and $\sigma_{yz} = \mu b x_M / 2\pi r^2$. The coordinates x_M, y_M and r refer to a point of the plane passing through M and perpendicular to oz. Indeed, this stress field is invariant by translations along Oz. Thus:

$$x_M = a, \qquad y_M = \frac{u}{\sqrt{2}}, \qquad r^2 = a^2 + (u^2/2) \tag{B.64}$$

and

$$\sigma_{xz} = \frac{-\mu b u}{2\sqrt{2}\pi(a^2 + u^2/2)} \qquad \sigma_{yz} = \frac{\mu b a}{2\sqrt{2}\pi(a^2 + u^2/2)} \tag{B.65}$$

The force exerted by D on the infinitesimal portion of dislocation du of center M is determined by the Peach-Koehler formula:

$$\overrightarrow{dF} = \overrightarrow{b'}.\overline{\overline{\sigma}} \wedge \overrightarrow{du} \tag{B.66}$$

Thus:

$$\overrightarrow{dF} = ((0, b'/\sqrt{2}, b'/\sqrt{2}) \begin{bmatrix} 0 & 0 & \sigma_{xz} \\ 0 & 0 & \sigma_{yz} \\ \sigma_{xz} & \sigma_{yz} & 0 \end{bmatrix} \wedge \frac{1}{\sqrt{2}} \begin{bmatrix} 0 \\ du \\ du \end{bmatrix} = \frac{b' du}{2} \begin{bmatrix} 0 \\ -\sigma_{xz} \\ \sigma_{xz} \end{bmatrix} \tag{B.67}$$

To prove the mentioned property it is sufficient to show that:

$$\int_{-\infty}^{+\infty} \sigma_{xz} du = 0 \tag{B.68}$$

or:

$$\int_{-\infty}^{+\infty} \frac{-\mu b u\, du}{2\sqrt{2}\pi(a^2 + u^2/2)} = 0 \tag{B.69}$$

Which vanishes by symmetry $(u \to u' = -u)$. One can also use the relation:

$$\int_{-\alpha}^{+\alpha} \frac{u\, du}{(a^2 + u^2/2)} = \left[Ln(a^2 + u^2/2) \right]_{-\alpha}^{+\alpha} = 0 \tag{B.70}$$

Exercise 15. *Interactions between dislocations*

1°) For the two screw dislocations, the Burgers vector are respectively $(0, 0, b)$ and $(0, b, 0)$. The stress field induced by $\overrightarrow{Oz}$ comprises σ_{13} and σ_{23} as only nonzero components. Let $\overrightarrow{dL} = \overrightarrow{j}\, dy$ be a segment of infinitesimal length of the dislocation $\overrightarrow{AY}$. The Peach-Koehler formula determines a force $\overrightarrow{dF}$ exerted on it:

$$\overrightarrow{dF} = (0, b, 0) \begin{pmatrix} 0 & 0 & \sigma_{13} \\ 0 & 0 & \sigma_{23} \\ \sigma_{13} & \sigma_{23} & 0 \end{pmatrix} \wedge \begin{pmatrix} 0 \\ dy \\ 0 \end{pmatrix} = \begin{pmatrix} -b\sigma_{23}dy \\ 0 \\ 0 \end{pmatrix} \tag{B.71}$$

with:

$$dF_x = -b\sigma_{23}dy = -bdy\frac{\mu bx}{2\pi r^2} = \frac{\mu b^2 a}{2\pi(a^2 + y^2)}dy \qquad (B.72)$$

The forces being equal for the two half axes $\pm y$, no torque is exerted on $\overrightarrow{AY}$. The sum of the forces is:

$$F_x = -\frac{\mu b^2}{2\pi}\int_{-\infty}^{+\infty}\frac{d(y/a)}{1 + (y/a)^2} = -\frac{\mu b^2}{2} \qquad (B.73)$$

Its sign shows that it is a force of mutual attraction.

2°) The stress field is the same as in the preceding case, as well as the segment $\overrightarrow{dL}$. However, the Burgers vector of the second dislocation is $(0, 0, b)$. Thus:

$$\overrightarrow{dF} = (0,0,b)\begin{pmatrix} 0 & 0 & \sigma_{13} \\ 0 & 0 & \sigma_{23} \\ \sigma_{13} & \sigma_{23} & 0 \end{pmatrix} \wedge \begin{pmatrix} 0 \\ dy \\ 0 \end{pmatrix} = \begin{pmatrix} 0 \\ 0 \\ b\sigma_{13}dy \end{pmatrix} \qquad (B.74)$$

and:

$$dF_z = -\frac{\mu b^2 ydy}{2\pi(a^2 + y^2)} \qquad (B.75)$$

The sum of these forces over the dislocation line is equal to zero since the forces exerted on $+y$ and $-y$ are opposite. The momentum is non-zero. The sum on each half-line is infinite. hence an infinite torque is exerted which tends to rotate $\overrightarrow{AY}$ towards $-\overrightarrow{z}$. the interaction tends to bring the two dislocations to antiparallel directions. In this final stage, the two dislocations, one edge-type and the other screw-type, are parallel and do not interact (cf. chapter 6).

Exercise 16. *Interaction of a dislocation with a surface*

1°) a) If z is parallel to the dislocation, the displacement (parallel to z) induced by the dislocation is (cf. chapter 6) $\xi = (b/2\pi)\theta$. The non-zero strain components are two shear components:

$$\epsilon_{13} = \epsilon_{xz} = -\frac{b}{4\pi r}\sin\theta = -\frac{b}{4\pi}\frac{y}{r^2} \qquad (B.76)$$

$$\epsilon_{23} = \epsilon_{yz} = \frac{b}{4\pi}\cos\theta = \frac{b}{4\pi}\frac{x}{r^2} \qquad (B.77)$$

In polar coordinates there is a single non-zero component $\epsilon_{\theta z} = b/4\pi r$. Likewise the stress components related to these strain components in an isotropic elastic medium are $\sigma_{13} = 2\mu\epsilon_{13}$, $\sigma_{23} = 2\mu\epsilon_{23}$ and $\sigma_{\theta z} = \mu b/2\pi r$. For $b > 0$ one has a helix which pointing towards increasing values of θ.

b) The stress field induced by the dislocation in the medium bordered by the free surface is the same as the field induced by two dislocations situated in

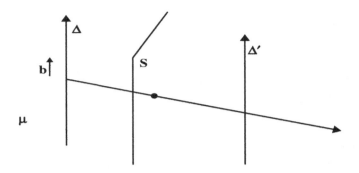

Figure B.13

an infinite medium. This can be established by checking that the equilibrium equation and the limit conditions are both satisfied. The equilibrium equation $2\mu\Delta\overrightarrow{\xi} + (\lambda + \mu)\overrightarrow{\nabla}(\mathrm{div}\overrightarrow{u\xi}) = 0$, being satisfied by the displacements induced by each of the dislocations, it is also satisfied by their sum. This is also the case for the vanishing at $\pm\infty$. It remains to check the condition $\overline{\overline{\sigma}}.\overrightarrow{n} = 0$ on the surface S. Let the x-axis be the line, normal to S, which joins the initially considered dislocation to its image. Let the y axis form a direct frame with the x and z axes. In cartesian coordinates, one can write:

$$\sigma_{13} = -\frac{\mu b}{2\pi r}\sin\theta - \frac{\mu b'}{2\pi r}\sin(\pi - \theta) \tag{B.78}$$

$$\sigma_{23} = \frac{\mu b}{2\pi r}\cos\theta + \frac{\mu b'}{2\pi r}\cos(\pi - \theta) \tag{B.79}$$

The force exerted on the surface is:

$$\overline{\overline{\sigma}}.\overrightarrow{n} = \begin{bmatrix} 0 & 0 & \sigma_{13} \\ 0 & 0 & \sigma_{23} \\ \sigma_{13} & \sigma_{23} & 0 \end{bmatrix} \begin{pmatrix} 1 \\ 0 \\ 0 \end{pmatrix} = \begin{pmatrix} 0 \\ 0 \\ \sigma_{13} \end{pmatrix} = 0 \tag{B.80}$$

The condition $\sigma_{13} = 0$ will be satisfied if $b' = -b$. Note that the stress σ_{23} does not vanish on the surface, but it does not exert any force on it.

c) This force is the same as that exerted by the "image dislocation". It is therefore attractive. The dislocation is attracted towards the surface and tends to migrate in its direction. The value of the force is that exerted between two screw dislocations (Eq. (6.1) of chapter 6). It is also possible to calculate this force by using the stress field induced by the "image" dislocation at the site of the "real" dislocation. This field is, (with $r = 2l$ and $\theta = \pi$)

$$\sigma'_{13} = 0 \quad \sigma'_{23} = \frac{\mu b}{4\pi l} \tag{B.81}$$

Applying the Peach-Koehler formula, the force per unit length is:

$$\vec{f} = \vec{b}\,\overline{\overline{\sigma}}\wedge\vec{\Delta} \tag{B.82}$$

where $\vec{\Delta}$ is the unit vector parallel to the dislocation (parallel to the z-axis). Thus:

$$(0,0,b)\begin{pmatrix} 0 & 0 & 0 \\ 0 & 0 & \sigma_{23} \\ 0 & \sigma_{23} & 0 \end{pmatrix} \wedge \begin{pmatrix} 0 \\ 0 \\ 1 \end{pmatrix} = \begin{pmatrix} b\sigma_{23} \\ 0 \\ 0 \end{pmatrix} = \frac{\mu b^2}{4\pi l}\vec{i} \tag{B.83}$$

d) The elastic energy u per unit length of the screw dislocation is provided by Eq. (5.9) (chapter 5):

$$u_{screw} = \frac{\mu b^2}{4\pi} Ln(\frac{l}{r_0}) \tag{B.84}$$

Assume that, beyond the distance l between the dislocation and the surface, the stress vanishes and that, therefore, the elastic energy is equal to zero. If this distance is decreased, the energy also decreases. Hence, the surface and the dislocation wil attract eachother in order to decrease their distance. The force is determined by $f = -\partial u/\partial l$:

$$|f_{screw}| = \frac{\mu b^2}{4\pi l} \tag{B.85}$$

(sign of f : if the force exerted on the dislocation is considered, the origin of l should be taken on the surface. l is then negative. Its algebraic value is $f = -\partial u/\partial(-l) > 0$. The force is thus directed towards the direction $x > 0$).

e) The preceding argument based on consideration of the energy allows, without calculation, to predict the attractive nature of the interaction, since the formula providing the energy of an edge dislocation (cf. Eq. (6.16)) has the same form as that corresponding to a screw dislocation. One can write $u_{edge} = u_{screw}/(1-v)$. One deduces that $f_{edge} = f_{screw}/(1-v)$.

f) Any parallel portion of a dislocation, parallel to the free surface, will tend, through their mutual attraction, to get closer to the surface and eventually to be evacuated through it. Hence, one expects that close to the surface the dislocations will be normal to the surface.

2°) a) In the medium with modulus μ, the sum of the normal stress components σ_{13} induced by Δ and Δ' on the surface yields:

$$\sigma_{13} = -\frac{\mu}{2\pi l}[b\sin\theta + b'\sin(\pi - \theta)] = -\frac{2\mu\mu' b\sin\theta}{2\pi l(\mu + \mu')} \tag{B.86}$$

Likewise, in the μ' medium, Δ'' détermines on the surface the same stress:

$$\sigma_{13} = \frac{-\mu'b''}{2\pi l}\sin\theta = -\frac{2\mu\mu'b\sin\theta}{2\pi l(\mu+\mu')} \tag{B.87}$$

Hence, in the μ medium, the sign of Δ' is the same as that of $(\mu'-\mu)$. There will be repulsion of Δ by the interface if $\mu' > \mu$ (μ' "stiffer" medium) and attraction if $\mu' < \mu$ (μ' "softer" medium). The dislocation tends to get more distant of a rigid medium in which its elastic energy (proportional to μ) would be larger.

b) The perfectly rigid surface can be assimilated to a medium with infinite value of μ'. The preceding result then leads to $b'' = 0$, that is to the non-existence of Δ''. Also: $b' = b$. Since the real dislocation Δ and the fictive one Δ' repell eachother, one can conclude that the rigid surface repells the dislocation Δ. This is in agreement with the fact that, the energy of a dislocation being proportional to μ, the dislocation will not tend to migrate towards the surface (whose μ value is infinite). The argument given in $1°$) d) is not valid. Indeed the stress does not vanish on the surface, and therefore, the energy of the dislocation cannot be evaluated for a limited volume of radius l.

$3°$) a) From the result in $1°$) b), a pair of dislocations of opposite signs, "images" of one-another in a surface, will comply with the limit conditions on this surface (Fig. B.14). Thus, a single pair $(\Delta^+ + A^-)$ is sufficient to have a zero normal stress on S_2. However, this pair determines in the medium μ a mechanical state corresponding to the presence of two dislocations (Δ^+ and A^-). To comply with the limit conditions on S_1 it is then necessary to add two "image" dislocations (A'^- and B^+) of signs respectively opposite to those of Δ and A^-. The mechanical state in the layer, which results from the presence of two additional dislocations (A'^+ and B^-) is such as to destroy the compatibility with the limit conditions on S_2. One has again to introduce two image-dislocations with respect to S_2, and so forth... Eventually, the mechanical state within the layer is the same as that of an infinite medium in which are present an infinite number of pairs of dislocations of opposite sign.

b) A dislocation $(\pm b)$ situated at the *right* of Δ exerts the stress (Eq. B.81) $(\mp\mu b/2\pi R)$, in which R is the distance between Δ and the dislocation. Likewise, a dislocation $(\pm b)$ situated at the *left* of Δ exerts a stress $(\pm\mu b/2\pi R')$. For the dislocations situated at the right, and of negative sign, the successive distances are $2l, 2a + 2l, 4a + 2l,...,2na + 2l, ...$The dislocations of positive sign are located at $2a, 4a, 6a, ...2na$. At the left, the negative dislocations are at the distances $2a - 2l, 4a - 2l, ...2na - 2l$, while the positive ones are at $2a, 4a, ...2na, ...$ Finally the sum of the contributions is:

$$\sigma_{23} = \frac{\mu b}{2\pi}\left[\sum_0^\infty \frac{1}{2na+2l} - \sum_1^\infty \frac{1}{2na} - \sum_1^\infty \frac{1}{2na-2l} + \sum_1^\infty \frac{1}{2na}\right] \tag{B.88}$$

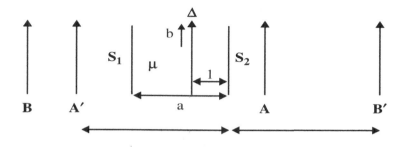

Figure B.14

Two of the sums in the bracket mutually cancel, and, in addition, one can write:

$$-\sum_{1}^{\infty} \frac{1}{2na - 2l} = \sum_{-\infty}^{-1} \frac{1}{2na + 2l} \tag{B.89}$$

The expression is then:

$$\sigma_{23} = \frac{\mu b}{2\pi} \sum_{-\infty}^{+\infty} \frac{1}{2na + 2l} \tag{B.90}$$

c) It is indicated that $\sigma_{23} = (\mu b/4\pi a)\cot g(\pi l/a)$. Equation (B.83) yields :

$$\vec{f} = b\sigma_{23}\,\vec{i} = \frac{\mu b^2}{4\pi a}\cot g\frac{\pi l}{a}.\,\vec{i} \tag{B.91}$$

Hence, the force is positive when $l < (a/2)$ and negative when $l > (a/2)$. It vanishes when $l = (a/2)$. Consistent with the qualitative considerations mentioned above, the dislocation is attracted towards the closest surface.

Exercise 17. *Dislocation in a cylindrical sample*

1°) The non-zero cartesian coordinates induced in an infinite solid medium by a screw dislocation are (cf. chapter 6):

$$\sigma_{13} = \frac{-\mu b\sin\theta}{2\pi r} \quad ;\sigma_{23} = \frac{\mu b\cos\theta}{2\pi r} \tag{B.92}$$

At point M of the surface, the normal pointing out of the sample is $\vec{n} = (\cos\theta, \sin\theta)$. The force exerted on an infinitesimal element dS of the surface is:

$$\vec{df} = \overline{\overline{\sigma}}.\vec{n}.dS = dS.[\sigma_{13}\cos\theta + \sigma_{23}\sin\theta] = 0 \tag{B.93}$$

One can therefore check that the limit conditions on the surface are satisfied by the above stress field defined for an infinite medium. On the other hand, the equations governing locally the elastic equilibrium of the solid are also satisfied. Hence, the stress field (B.92) is the equlibrium field in the cylinder.

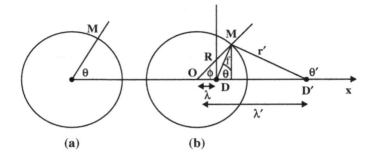

Figure B.15

2°) Referring to Fig. B.15 one can write:

$$r \sin \theta = R \sin \phi$$

$$r^2 = R^2 + \lambda^2 - 2R\lambda \cos \phi$$

$$r \cos \theta = R \cos \phi - \lambda$$

hence the form of the cartesian components (B.92) as function of R, λ, ϕ :

$$\sigma_{13} = \frac{-\mu b}{2\pi} \frac{R \sin \phi}{R^2 + \lambda^2 - 2R\lambda \cos \phi}$$

$$\sigma_{23} = \frac{\mu b}{2\pi} \frac{R \cos \phi - \lambda}{R^2 + \lambda^2 - 2R\lambda \cos \phi}$$

3°) The force is provided by Eq. (B.93) by replacing θ by ϕ, thus:

$$df = \frac{-\mu b \lambda \sin \phi}{2\pi (R^2 + \lambda^2 - 2R\lambda \cos \phi)}$$

This force does not vanish on the entire surface. The preceding stress field is therefore not compatible with the limit conditions, imposing a zero force exerted on the surface, in contrast to the situation in 1°).

4°) The stress field determined by D' has the same expression as that determined for D up to a substitution of $(-b)$ to (b) and of λ' to λ. Hence the sum of the forces exerted on the surface is

$$df = \frac{-\mu b \lambda \sin \phi}{2\pi(R^2 + \lambda^2 - 2R\lambda \cos \phi)} + \frac{\mu b \lambda' \sin \phi}{2\pi(R^2 + \lambda'^2 - 2R\lambda' \cos \phi)}$$

Writing the vanishing of the force leads to:

$$\lambda'((R^2 + \lambda^2 - 2R\lambda \cos \phi) = \lambda(R^2 + \lambda'^2 - 2R\lambda' \cos \phi)$$

or:

$$\lambda(\lambda')^2 - \lambda'(\lambda^2 + R^2) + \lambda R^2 = 0$$

The solutions of this equation are $\lambda' = \lambda$ and $\lambda' = (R^2/\lambda)$. The first solution correponds to the absence of a dislocation (annihilation of two dislocations with opposite Burgers vectors). The second solution is the suitable one.

5°) a) For a screw dislocation, any plane containing the dislocation is a glide plane. This is the case of (D, Ox).

b) D is submitted to the stress field exerted by the surface, which is the same as the one exerted by D' inside the cylinder. As D' and D have opposite signs, the surface exerts an attractive force, whose modulus per unit length of the dislocation is (cf. chapter 6):

$$f = \frac{\mu b^2}{2\pi d} = \frac{\mu b^2 \lambda}{R^2 - \lambda^2} \tag{B.94}$$

This force will attract the dislocation towards the surface and lead to its evacuation from the sample.

c) The answer is negative, because the step produced is parallel to b. It would be observable on a surface normal to the axis of the cylinder.

Exercise 18. *Energy of the dislocation-core*

1°) In the framework of the continuum description (cf. chapter 7), in which $[\xi(x_{n+1}) - \xi(x_n)] = \xi'(x_n)(x_{n+1} - x_n) = b(d\xi/dx)$, the elastic-part of the energy referred to the infinitesimal interval dx can be written as (there are (dx/b) terms in the discrete sum):

$$\sum \rightarrow \int \frac{dx}{b} \rightarrow U_{elast} = \int Kb(\frac{d\xi}{dx})^2 dx \tag{B.95}$$

On the other hand, assuming the validity of the expression relative to an elastic medium, and taking into account that $(\sigma_{ij}\epsilon_{ij} + \sigma_{ji}\epsilon_{ij})$: $(1/2)[\sigma\epsilon + \sigma\epsilon]$, the misfit-part of the energy referred to the unit volume, can be written as $(\sigma^2/2\mu)$. Thus the misfit-energy of the core per length b of the dislocation (volume b^3 per atome, (dx/b) atoms along the glide plane) is:

$$U_{misfit} = \int \frac{\mu b^2}{8\pi^2} \sin^2(\frac{4\pi\xi}{b})dx \tag{B.96}$$

Note that the expressions found for the two contributions are dimensionally consistent. Indeed (U: energy, L: length : $[K][b][\xi'^2]=[U][L]^{-2}[L]=[U][L]^{-1}$ and $[\mu b^2]=[UL^{-3}L^2]$).

2°) Replacing $(d\xi/dx)^2$ by its expression as function of $\cos(2\pi\xi/b)$ (cf. chapter 7), the total energy is:

$$U_c = \frac{\mu b^2}{2\pi^2} \int_{-\infty}^{+\infty} \cos^2(\frac{2\pi\xi}{b})[1 + \sin^2(\frac{2\pi\xi}{b})]dx \tag{B.97}$$

Putting in U_c the expression of $\xi(x)$, one obtains, with $t = \tan(\pi\xi/b) = -th(2x/w)$:

$$U_c = \frac{\mu b^2}{2\pi^2} \int_{-\infty}^{+\infty} f\{th(2x/w)\}dx \tag{B.98}$$

which is the required expression with $u = 2x/w$.

3°) $U_c < 2\mu b^2 w/4\pi^2$. w being in the range $2b$-$4b$, one obtains $U_c < \mu b^3/5$ for a dislocation length equal to b (assuming a section equal to the area of one unit-cell). Per unit length, the core energy is $U < \mu b^2/5$. As a comparison, the elastic energy induced by an edge dislocation in the "good crystal" (cf. chapter 6) is:

$$\frac{\mu b^2}{4\pi(1 - \nu)} Ln\frac{R}{r_0} \tag{B.99}$$

With $\nu \approx 0.5$, $R \approx 100$ microns and $r_0 = w \approx 2b$, one obtains $U_{elast} \approx 2\mu b^2 > 10 U_c$. The core energy therefore contributes to an amount of 10% to the energy of the dislocation.

Exercise 19. *Dissociation of the dislocation core*

1°) a) These are the primitive translations perpendicular to (111), hence $(1/2)(\vec{c}_i - \vec{c}_j)$

b) It is the projection on the (111) plane of a vector joining the atomic positions of the B plane and of the A plane. Several choices are possible. The appropriate vector is, for instance (see Fig. A.18):

$$\vec{b} = \frac{1}{3}[\frac{1}{2}(\vec{c}_2 - \vec{c}_1) + \frac{1}{2}(\vec{c}_2 - \vec{c}_3)] = \frac{1}{6}(2\vec{c}_2 - \vec{c}_1 - \vec{c}_3) \tag{B.100}$$

$\vec{b}$ makes an angle of 30° with the (horizontal) dislocation line D. This line is therefore of the mixed type. The Burgers vector of the screw component is the horizontal projection of $\vec{b}$: $b\sqrt{3}/2$.(written in vectorial form this component is $\vec{b}_s = \frac{1}{4}(\vec{c}_2 - \vec{c}_1)$). The Burgers vector of the edge component is the vertical projection of $\vec{b}$, thus $b/2$ (vector form $\vec{b}_e = \frac{1}{12}(\vec{c}_2 + \vec{c}_1 - 2\vec{c}_3)$).

c) $\vec{b}$ is not a vector of the Bravais lattice (it is not a linear combination with integer coefficients of the $\frac{1}{2}(\vec{c}_i - \vec{c}_j)$). D is therefore not an ordinary dislocation (i.e. a linear defect). Above D, (Fig. A.18) one has an abnormal sequence AC between adjacent planes, hence a half-plane of defects, which is a "stacking fault" (see chapters 2 and 7). D is the edge of a stacking fault. It can be considered as a dislocation, in a broad sense, as it is a line of discontinuity of the displacement corresponding to a uniform displacement $\vec{b}$ across the line (ignoring the transition region of the core).

2°) a) It is an imperfect dislocation since $\vec{b}'$ is not a Bravais lattice translation. Its screw component is $\vec{b}'_S = \frac{1}{4}(\vec{c}_2 - \vec{c}_1)$ and its edge component is $\vec{b}'_E = \frac{1}{12}(2\vec{c}_3 - \vec{c}_2 - \vec{c}_1) = -\vec{b}_e$ (Burgers vector of the edge component of D). If one starts from an atomic position in plane B, the two succesive shifts $\vec{b}$ and $\vec{b}'$ lead to a total shift equal to:

$$\vec{b} + \vec{b}' = \frac{1}{2}(\vec{c}_2 - \vec{c}_1) \tag{B.101}$$

which is a primitive translation in the B plane. The final atomic position is therefore a standard atomic position of the B-type.

b) When D and D' coincide, the upper part of the figure can be considered to have slipped by $(\vec{b} + \vec{b}')$ with respect to the lower part. This vector being a primitive translation of the lattice, parallel to the glide plane, the line of discontinuity is a perfect dislocation. Its Burgers vector being parallel to the line, the dislocation is of the screw-type.

3°) a) The screw and edge portions of the dislocations interact separately (cf. chapter 6). The two screw components have the same Burgers vector. Par unit length, they repell eachother with the force:

$$\mu b_s^2 / 2\pi d = \frac{\mu c^2}{16\pi d} \tag{B.102}$$

in which c is the edge of the cubic cell. The two edge components, which share the same glide plane ($\theta = 0$ in formula (6.31) of chapter 6) have opposite Burgers vectors. They attract eachother with the force:

$$\frac{\mu b_c^2}{2\pi(1-\nu)d} = \frac{\mu c^2}{48\pi(1-\nu)d} \tag{B.103}$$

The total force is repulsive (for $\nu = 0, 5$), and equal to :

$$F = \frac{\mu c^2}{48\pi d} \tag{B.104}$$

b) The energy of the band comprised between D and D' is γd, per unit length of the dislocations. If a dislocation is shifted by δd this energy varies by an amount $\gamma.\delta d$. It can thus be deduced that, between the dislocations, there is an effective attractive force equal to γ per unit length.

c) This distance is determined by the equality between the two forces.

$$\gamma = \frac{\mu c^2}{48\pi d_{eq}} \rightarrow d_{eq} = \frac{\mu c^2}{48\pi\gamma} \tag{B.105}$$

d) Expressing all the quantities in meters and in Joules, one has $\mu = 7.5.10^{10}$ $c^2 = 13.10^{-20}m^2$; $d_{eq} = 5.10^{-9}m$. Thus:

$$\gamma \approx \frac{(7.5).13}{48.5}.10^{-1} J/m^2 \approx 4.10^{-2} J/m^2 \tag{B.106}$$

e) One has $E_{total} = 2\gamma.d_{eq} = \frac{\mu c^2}{24\pi} = 2.10^{-10} J/m$. The energy of a perfect dislocation has been evaluated in the framework of the Volterra description (chapters 5 and 6). In the framework of the same description, the expression remains valid for an imperfect dislocation. One can therefore write for the set of two imperfect dislocations:

$$U = \frac{\mu}{4\pi}(2b_v^2 + \frac{2}{1-\nu}b_c^2)Ln(\frac{R}{r_c}) = \frac{\mu}{4\pi}(\frac{c^2}{4} + \frac{c^2}{6})Ln(\frac{R}{r_c}) \approx \frac{\mu c^2}{2\pi} \tag{B.107}$$

The interaction energy is much smaller than this sum. The total energy is $(13\mu c^2/24\pi)$

4°) The energy of a perfect dislocation is thus reduced to its Volterra contribution:

$$\frac{\mu}{4\pi}b_0^2 Ln(\frac{R}{r_c}) \approx \frac{5\mu c^2}{8\pi} = \frac{15\mu c^2}{24\pi} \tag{B.108}$$

in which $b_0 = (c^2/2)$ is the Burgers vector of the perfect dislocation. This value is larger than that calculated for the two imperfect dislocations D and D'. The perfect dislocation is therefore not energetically favoured.

Exercise 20. *Dissociation of a dislocation*

1°) A Burgers circuit surrounding the two dislocations must determine the same closure defect $\vec{b}$ than the initial dislocation. One can write therefore $\vec{b} = \vec{b}_1 + \vec{b}_2$ and thus, $\vec{b}_2 = (c/2)(\vec{i} + \vec{j} - \vec{k})$. These two vectors are primitive translations of the crystal. The elastic energy U associated to the initial dislocation is proportional to $\vec{b}^2$ while the energy U_{12} associated to the dislocations $\vec{b}_i$ is proportional to $(\vec{b}_1^2 + \vec{b}_2^2)$. One can write:

$$U \propto \vec{b}^2 = 2c^2$$

and

$$U_{12} \propto (\vec{b}_1^2 + \vec{b}_2^2) = 3c^2/2 < U$$

The initial dislocation is thus unstable with respect to the considered decomposition.

2°) a) When the ineteraction between D_1 and D_2 is neglected, the elastic energy'of the two dislocations is proportional to $(\vec{b}_1^2 + \vec{b}_2^2)$, with $\vec{b}_1 = \frac{c}{6}(2\vec{i} - \vec{j} - \vec{k})$ and $\vec{b}_2 = (\vec{b} - \vec{b}_1) = \frac{c}{6}(\vec{i} - 2\vec{j} + \vec{k})$, thus:

$$U_{12} \propto \frac{c^2}{3} < U_D \propto \frac{c^2}{2}$$

D has therefore a tendency to dissociate.

b) For two screw dislocations (chapter 6, Eq. (6.30)):

$$F_{12} = \frac{\mu \vec{b}'_1 \vec{b}'_2}{2\pi\delta} = +\frac{\mu c^2}{24\pi\delta}$$

The interaction is repulsive, since $\vec{b}'_1 \vec{b}'_2 > 0$. Such a conclusion could be expected because, if the two dislocations are close to each other, their energy is equal to that, U_D, of D, while if they are far from each other, this energy is $U_{12} < U_D$. The energy decreases when their distance increses, and, thus they repell each other.

c) The dislocations have Burgers vectors which are not Bravais lattice translations (In the FCC-lattice the primitive translations are of the type $\pm(\vec{i} \pm \vec{j}), \pm(\vec{k} \pm \vec{j}), \pm(\vec{i} \pm \vec{k})$) and the $\vec{b}_i$ are not linear combinations of the preceding vectors.

As a consequence, the area of the portion of plane which separates the two dislocations, has slipped by a translation distinct from a lattice translation. It is a defective area.

d) Per unit length of the two dislocations, the area of the plane P is δ. The energy per unit length of the considered fraction of P is thus $U_P = f\delta$. The force of the interaction is $-(dU_P/d\delta) = -f$. The equilibrium distance corresponds to the equality between this attractive force and the repulsive force calculated in b). It provides the value:

$$\delta_0 = +\frac{\mu c^2}{24\pi f}$$

e) σ is the Peierls-Nabarro stress $\sigma_{PN} = 2\mu \exp(-\gamma W/b)$ in which W is the width of the core of the dislocation and in which γ is a coefficient whose value is of the order of π. The numerical application yields $(\delta_0/b') = \delta_0\sqrt{2}/c \approx 4.7$. Hence, assuming $\gamma = \pi$ (Another choice, e.g. 2π is considered also correct), $\sigma = 10^{-4}\mu$, in agreement with the experimental values currently observed.

Exercice 21. *Deformation of a rod*

1°) The most favourable glide planes are, as shown in exercise 2, the hexagonal compact planes perpendicular to the c-axis. The possible glide directions in these planes are those for which the atomic spacings are smallest. They are also directions of the possible Burgers vectors $\vec{b}$, which are the shortest translations of the lattice in these planes i.e. $\pm\vec{a}_1; \pm\vec{a}_2; \pm(\vec{a}_1 - \vec{a}_2)$.

2°) a) The dislocations density being uniform, and the rod being $1cm$ in diameter, there are approximately 10^4 dislocations uniformly distributed on a height of $1cm$ parallel to c.In a hexagonal compact structure, the spacing between planes in which the dislocations are likely to be localized, is $c/2 = a\sqrt{2/3} \approx$ 2Å. For each $1cm$ height of the rod, there are therefore 5.10^7 planes, and, an average of 2.10^{-8} dislocation per plane (which means, that, in a plane, there is, in general, zero or one dislocation). Hence, the average distance between planes which contain a dislocation is $\approx 10^{-4}cm = 10^4$Å.

b) The observed spacing between planes corresponds to the fact that only the compact planes containing a dislocation are likely to contain a dislocation source. Their spacing calculated in a) agrees with the observed value of the spacing.

c) The displacement of a single dislocation can lead, at most, to a relative gliding equal to $b = 2.5$Å. The observation of glide-amplitudes of the order of 10^3Å along certain planes means that, in these planes, there is a source (or several sources) of dislocations (e.g. of the Frank-Read type). In turn, such a source denotes the occurence of a pinning of mobile dislocations at definite points (e.g. by interaction with the "forest" of dislocations, or with sessile dislocations). The functioning of the sources generates mobile dislocations continuously and explains the occurence of a large value of the glide-amplitude (here 10^3Å=400 dislocations evacuated from the sample).

Exercise 22. *Elastic limit*

metal→	beryllium	Cobalt	Zinc
c/a	1.59	1.62	1.86

(B.109)

The Peierls-Nabarro stress is determined by:

$$\sigma_{PN} = 2\mu e^{-\gamma W/b} \tag{B.110}$$

In a hexagonal compact metal, The Burgers vector b in the glide direction, within the compact planes, is equal to the modulus of the primitive translations a in these planes. On the other hand, the width W of the dislocation core is equal:

$$W = 2\sqrt{\frac{2\alpha K}{\mu}} \tag{B.111}$$

in which α is the distance between atomic planes, normal to the glide planes (also equal to the primitive translation in this direction). These are hexagonal compact planes. One has therefore $\alpha = c$. The constant K controls the restoring force between atoms, parallel to the glide plane. It is likely that K has the same order of magnitude in all the preceding structures which are formed by a stacking of compact planes in which the atoms are in contact with each other.

The easiest the shearing of adjacent glide planes, the smaller the value of μ. In agreement with the discussion in section 7.2.3, this will be the case if the spacing between adjacent planes is large. The preceding considerations finally lead to the conclusion that W (measured in units $b = a$) will be larger if (c/a) is larger. Hence, it will determine a smaller value of σ_{PN}. The yield stress being assumed to be equal to σ_{PN}, the required classification will be:

$$\sigma_{PN}^{zinc} < \sigma_{PN}^{cobalt} < \sigma_{PN}^{beryllium} \tag{B.112}$$

Zinc is the "softest" of the considered metals, and beryllium the "hardest".

Exercise 23. *Yield stress*

1°) In the absence of dislocations, the yield stress is equal to the ideal shear stress σ_c relative to the direction of planes of the slips (portion OA of the strain-stress curve, A being the yield stress). When such a stress is applied, dislocations are generated (since partial slips will occur). The value of the stress which induces permanent deformations is then the stress required to move the existing dislocations. It is equal to the Peierls-Nabarro stress, whose value is generally several orders of magnitude smaller than the ideal shear stress. Hence, the large decrease of σ (Portion AB of the curve). The remaining part of the curve (BC) corresponds to a moderate hardening. The yield stress σ_A is equal to $\alpha . \sigma_c$ in which α depends of the orientation of the applied stress with respect to the glide planes (Schmid factor). The answer $\sigma_A = \sigma_c$, would also be correct. On the other hand, $\sigma_B \approx \sigma_{PN}$: Peierls-Nabarro stress.

2°) a) In regard of the detailed mechanism of a Frank-Read source (chapter 8), figure B.16 constitutes a clear answer: the beginning of the sequence of shapes is identical to that expected for a Frank–Read source, i.e. increasing curvature induced by the balance between the line tension and the forces, normal to D, produced by the external stress; stage corresponding to a half-circle limited by two precipitates, then free expansion. By contrast to the standard scheme, the stage of formation of successive loops is offset. Indeed, the dislocation arches fixed on neighbouring precipitates, give rise to a mutual annihilation of segments facing eachother in two adjacent arches. A quasi-straight dislocation is eventually obtained, as well as small loops surrounding the precipitates. The straight part moves to the right and encounters another column of precipitates which is crossed by the same mechanism.

b) The yield stress is the stress required to overcome the obstacles constituted by the precipitates. This is the shear stress required to form a semi-circular dislocation pinned by two neighbouring precipitates. The formula in chapter 8 relative to the Frank-Read source can be used in this view:

$$\sigma_L = \frac{2\alpha\mu b}{a} \tag{B.113}$$

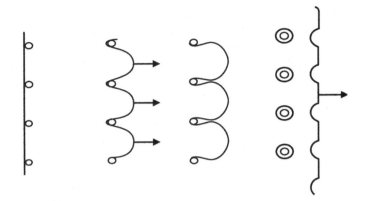

Figure B.16

with

$$\alpha = \frac{1}{4\pi(1-\nu)} Ln(\frac{A}{W/2}) \qquad (B.114)$$

since the distance at which the stress field of the dislocation vanishes, is equal to A (denoted R in formula (5.27) of chapter 5).

c) The remaining traces of the passage of the dislocation are the small loops surrounding the precipitates.

d) These loops induce an energy per unit length (ignoring their edge- or screw-nature):

$$\frac{\mu b^2}{4\pi} Ln(\frac{r}{W/2}) \qquad (B.115)$$

As this energy is confined to a torus of radius r, one can be consider that half of this energy is localized within the precipitate and the other half outside the loop. The approximate length of the loop is $2\pi r$. The increase of elastic energy of the precipitate is therefore approximately:

$$U_p \approx \frac{\mu b^2 r}{4} Ln(\frac{r}{W/2}) \qquad (B.116)$$

Exercise 24. *Reaction between dislocations*

1°) The two dislocations (Fig. B.17) are of mixed-type, since $\vec{b}$ makes an angle of 45° with their direction. the glide plane of AB, defined by AB and by $\vec{b}$ is the vertical plane represented on the figure. Likewise, the glide plane of CD is the horizontal plane containing this dislocation.

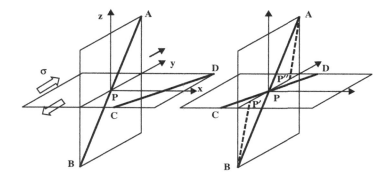

Figure B.17

2°) a) The line tension of a dislocation is $T = C\mu b^2$ in which, for a straight dislocation of mixed-type, C is determined by Eq. (5.17) of chapter 5:

$$C = \frac{1}{4\pi} Ln(\frac{R}{r_C})[\cos^2\theta + \frac{\sin^2\theta}{1-\nu}] \tag{B.117}$$

θ is the angle of $\vec{b}$ with the dislocation line. $\theta = \pi/4$ for the two dislocations. Hence:

$$T_{AB} = T_{CD} = \frac{\mu b^2}{4\pi} \cdot \frac{2-\nu}{2(1-\nu)} Ln(\frac{R}{r_C}) \tag{B.118}$$

b) The line tension has the effect of tending to shorten the length of the dislocation line and thus to make it straight.

3°) The calculation of the forces per unit length, for AB and CD, involves a common part, that of $\vec{V} = \vec{b}.\overline{\overline{\sigma}}$:

$$\vec{V} = (0b0) \begin{bmatrix} 0 & 0 & 0 \\ 0 & 0 & \sigma \\ 0 & \sigma & 0 \end{bmatrix} = \begin{pmatrix} 0 \\ 0 \\ b\sigma \end{pmatrix} \tag{B.119}$$

Then, respectively for the two dislocations the force is:

$$AB \rightarrow \begin{pmatrix} 0 \\ 0 \\ b\sigma \end{pmatrix} \wedge \frac{1}{\sqrt{2}} \begin{pmatrix} 0 \\ -1 \\ -1 \end{pmatrix} = \frac{1}{\sqrt{2}} \begin{pmatrix} b\sigma \\ 0 \\ 0 \end{pmatrix} \tag{B.120}$$

$$CD \rightarrow \begin{pmatrix} 0 \\ 0 \\ b\sigma \end{pmatrix} \wedge \frac{1}{\sqrt{2}} \begin{pmatrix} 1 \\ 1 \\ 0 \end{pmatrix} = \frac{1}{\sqrt{2}} \begin{pmatrix} -b\sigma \\ b\sigma \\ 0 \end{pmatrix} \tag{B.121}$$

The two forces are normal to the direction of the dislocation lines. Besides, the force parallel to x, exerted on AB, is normal to the glide plane of this dislocation (and directed towards $+x$), while the force exerted on CD is contained in its glide plane and directed to the left ($-x$ and $+y$).

4°) At ordinary temperatures, only the gliding of a dislocation is possible (not the climb). As the dislocation AB is pulled normal to its glide plane, it cannot move. By contrast CD is able to move towards $+y$, in such a way as to intersect the dislocation AB in P.

5°) a) It is consistent to consider CPB and APD because there is a continuity of orientations of these segments of dislocations, as well as a unique value of the Burgers vector $\vec{b}$ along them.

b) In the absence of stress σ, the line tension of each L-shaped dislocation will lead to a shortening of their length. This is achieved by displacing the points P' and P" in the indicated directions. One thus decreases the value of the elastic energy associated to the dislocations. The considered motions are displacements of the segments in their respective glide planes, and this is possible at ordinary temperatures.

6°) a) Using the Peach-Koehler formula applied to the dislocations directions $(0, \xi; \xi')$ (relative to the segments P'B and AP") and $(\xi, \xi', 0)$ (relative to the segments CP' and P"D) one can show that the forces remain normal to the glide plane of P'B and AP" and parallel to the glide planes of CP' and P"D.

b) The segments situated in the vertical plane are blocked while the segments situated in the horizontal plane can move. Since points P' and P" are fixed through the blocking of segments P'B and AP", the horizontal segments will give rise to spiral source as explained in section 8.4.3 of chapter 8.

c) The "sweeping", by the spiral source, of the surface of the sample, will generate relative slips, of amplitude b, of the portions of crystal situated on either sides of the horizontal plane. Such slips will induce a permanent (plastic) shear of the sample under the action of the stress σ_{yz}.

Exercise 25. *Dislocations source*

a) The convention defined in chapter 5 assigns, in a direct-oriented frame, the $\overrightarrow{Ox}$ axis to $\vec{b}$, the $\overrightarrow{Oz}$ axis to the oriented dislocation line and the $\overrightarrow{Oy}$ axis to the direction of the additional half-plane. Hence, this plane is situated at the left of the segment represented in Fig. A. 22.

b) The force exerted on the dislocation is assumed to be oriented to the right, thus towards $(-Oy)$. Consequently, with $F > 0$, one can write:

$$\begin{pmatrix} 0 \\ -F \\ 0 \end{pmatrix} = \begin{pmatrix} +b & 0 & 0 \end{pmatrix} \begin{pmatrix} \sigma_{xx} & \sigma_{xy} & \sigma_{xz} \\ \sigma_{xy} & \sigma_{yy} & \sigma_{yz} \\ \sigma_{xz} & \sigma_{yz} & \sigma_{zz} \end{pmatrix} \wedge \begin{pmatrix} 0 \\ 0 \\ +1 \end{pmatrix}$$

which provides $F = \sigma_{xx} > 0$. The stress is therefore a traction, normal to the plane of Fig. A. 22.

c) A forward motion of the dislocation extends the area of the additional half-plane. It is therefore a "climb" of the dislocation, which requires a diffusion of atoms at the left of the segment.

d) At each cycle of the source, the additional half-plane will occupy the entire area parallel to the plane represented in Fig. A. 22. Each cycle thus corresponds to the insertion of an additional plane in the solid. This is consistent with the fact that the applied traction, parallel to x, tends to lengthen the crystal in this direction.

Exercise 26. *Frank-Read double windmill*

$1°$) The configurations (b) et (c) are identical (simultaneous inversion of the sense of the left dislocation and of its Burgers vector). The sequence of shapes of the segment in each configuration is represented on figure B.18.

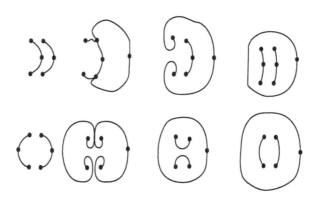

Figure B.18

$2°$) It is configuration (a) because the repulsion between dislocations exerts a stress on each dislocation which adds up to the external stress. The functioning of the source will therefore require a smaller external stress.

Exercise 27. *Dislocations energy*

a) The dislocations being essentially produced during the hardening process under action of the stress σ_{zz}, it can be assumed that they are generated by Frank-Read sources located in planes parallel to the xy plane and emitting dislocations, parallel to y, in the direction x.

b) The N dislocations present in the volume of the sample determine a density N/cm^2. The energy of a dislocation per unit length is:

$$\approx \frac{\mu b^2}{4\pi} Ln(\frac{R}{r_0})$$

The core radius is of the order of magnitude of b. One can choose $R = 1cm$ (dimension of the sample), or $1cm/\sqrt{N}$ equal to the average distance between dislocations. Either choice does not change the order of magnitude of the result since it is the logarithm $Ln(R/b)$ which appears in the expression of the energy. In the first case, $Ln(R/r_0) \approx 15 \approx 4\pi$. Hence, the dislocations being approximately $1cm$ long:

$$N\mu b^2 .10^{-2} = 10^2$$

and:

$$N = \frac{10^2 .10^2}{7.10^{10}.(2,5)^2.10^{-20}} \approx 2.10^{12}$$

Starting from the preceding number, the distance between dislocations can be evaluated. Thus, $Ln(R/r_0) \approx 4$, a value of N of the order of $6.10^{12}/cm^2$ is found.

Exercise 28. *Edge dislocation and vacancies*

1°) From equation (4.6) in chap. 4, this probability is:

$$q_0 = \frac{n_L}{N} = \exp(-\beta u_F) \tag{B.122}$$

2°) a) When a vacancy is introduced, the forces due to the pressure will produce a positive work if $p > 0$, and a negative work if $p < 0$. Another manner to describe the same situation is to emphasize that a positive pressure will tend to decrease by taking advantage of the decrease in volume associated to the vacancy. The elastic energy of the solid will therefore decrease by an amount equal to the variation of work. Hence, this energy will increase if $p < 0$.

b) $\delta u_F = -p\delta\Omega$. In order to determine the required probability, one can replace in eq.(B.122) u_F by $(u_F + \delta u_F)$. Referring to the signs of the energy changes which have been determined:

$$q = \exp[-\beta(u_F + p\delta\Omega)] \tag{B.123}$$

3°) a) Table 6.1 in chapter 6 shows the form of the components of the stress tensor induced by an edge dislocation. A non-zero pressure $p = -(1/3)\sum \sigma_{ii}$ is established everywhere in the solid. This pressure varies in space. Hence, the probability of presence of the vacancies (and relatedly, their concentration) varies, according to the results of 2°)b).

b) One can write:

$$p = -(1/3) \sum \sigma_{ii} = \frac{\mu b(1+\nu)\sin\theta}{3\pi(1-\nu)r}$$

Thus, applying eq.(B.123)

$$q(r,\theta) = \exp\left\{-\beta\left[u_F + \frac{\mu b(1+\nu)\sin\theta}{3\pi(1-\nu)r}\delta\Omega\right]\right\}$$

c) The pressure is positive for $0 < \theta < \pi$ and negative in the other half-plane. taking into account the (negative) sign of $\delta\Omega$, the probability of the vacancies is larger in the "compressed" region of the solid $(0 < \theta < \pi)$.

d)

$$\frac{q(3b, \pi/2)}{q_0} = \exp(\beta\frac{\mu b(1+\nu)}{3\pi(1-\nu)r}|\delta\Omega|) \approx \exp(\beta\mu b^3/62) \approx 15$$

Thus: $q(3b, \pi/2) = 15q_0 = 15\exp(-\beta u_F) = 15\exp(-10) = 6.6.10^{-4}$.

4°) a) A proportion of vacancies equal to q is realized on the edge of the additional atomic half-plane associated to the dislocation line. Considering the line of atoms constituting the edge of this half-plane, it appears that a vacancy corresponds to a pair of "jogs". Each jog has a length b and is parallel to Oy.

b) The probability of an upwards shift of amplitude b of the dislocation line is q. The average shift is qb. This shift of the dislocation line is a "climb".

c) Applying the Peach-Koehler formula shows that the force exerted on the Oz line, parallel to Oy tends to induce a climb of the dislocation line. Its effect on a pair of jogs is to increase the spacing between the two jogs, with the result of also inducing a climb of the dislocation. }

Exercise 29. *Dislocation climb*

1°) The Peach-Koehler formula détermines the force per unit length:

$$\vec{F} = (b,0,0)\begin{pmatrix} -\sigma & 0 & 0 \\ 0 & 0 & 0 \\ 0 & 0 & 0 \end{pmatrix} \wedge \begin{pmatrix} 0 \\ 0 \\ 1 \end{pmatrix} = \begin{pmatrix} 0 \\ b\sigma \\ 0 \end{pmatrix} \quad \text{(B.124)}$$

It is directed towards the additional half-plane, in a sense tending to reduce its extension. Its value is $b\sigma$.It cannot move the dislocation since such a motion would imply the diffusion of vacancies, and this is only possible at high enough temperatures.

2°)a) From Table 6.1 of chapter 6:

$$\sigma_{11} = -\frac{D\sin\theta(\cos 2\theta + 2)}{r}; \sigma_{22} = \frac{D\sin\theta\cos 2\theta}{r}; \sigma_{33} = \frac{-2D\nu\sin\theta}{r};$$

$$\sigma_{12} = \frac{D\cos\theta\cos 2\theta}{r} \quad \text{(B.125)}$$

with $D = 2\mu b / 4\pi(1 - \nu)$. Or $P = (-1/3) \sum \sigma_{kk}$ thus:

$$\Delta u = P \Delta V = \frac{\mu b(1 + \nu) \sin \theta}{3\pi(1 - \nu)r} \Delta V \tag{B.126}$$

at a point of polar coordinates (r, θ).

b) At thermal equilibrium, the atomic concentration of vacancies is:

$$c = e^{-\beta E} \tag{B.127}$$

in which E is the energy of a vacancy. In presence of a dislocation, $E = (u_F - \Delta u)$. Thus:

$$c(r, \theta) = \exp(-\beta u_f) \exp(\beta \Delta u) \tag{B.128}$$

the region of the solid which has the largest concentration of vacancies is therefore the "compressed" region ($\sin \theta = 1$) closest to the core:

$$c_{max} = c(R_0) = \exp(-\beta u_F) \exp(\beta \frac{\mu b(1 + \nu)}{3\pi(1 - \nu)R_0} \Delta V) \tag{B.129}$$

3°) When a vacancy reaches the dislocation, it substitutes an atom and forms a "jog" on the dislocation line. As a result, this line moves backward locally. A large number of vacancies will cause the entire line to move backwards: this motion is a "climb" of the dislocation.

4°) a) In a one-dimensional medium, and in a steady state regime, the diffusion equation which governs the evolution of the volume concentration reduces to:

$$\frac{\partial c}{\partial t} = 0 = -D_L \Delta c = -D_L \frac{d^2 c}{dr^2} \tag{B.130}$$

The variation law of c is thus linear. At $r = R_0$ its value is c_{max} while at $r = 0$ it is equal to zero, since D absorbs the vacancies and is therefore a vacancy "sink". Note that at R_0, if the atomic concentration is c, the volume concentration is c_{max}/b^3. One therefore obtains the expression indicated in the text:

$$c(r) = \frac{c_{max}}{b^3} \cdot \frac{r}{R_0} \tag{B.131}$$

b) The flow of vacancies is determined by the Fick equation (cf. Eq. (4.27)):

$$J = -D_L \frac{dc}{dr} = -\frac{D_L \cdot c_{max}}{R_0 b^3} \tag{B.132}$$

It is directed towards $(-\vec{r})$, hence towards the dislocation. Referred to a length b of dislocation, the number of vacancies per unit time is:

$$\frac{dn}{dt} = |J|.\pi R_0 b = \pi.D_L.b.c_{\max}/b^3 \tag{B.133}$$

c) To move backwards the dislocation by one interatomic spacing, a single vacancy must reach each length b of the dislocation line. If (dn/dt) vacancies reach this portion of the dislocation, it will move backwards by (dn/dt) interatomic spacings per unit time. The velocity of the motion is thus equal to (dn/dt).

d)

$$\mathbf{v} = \frac{dn}{dt} = \frac{\pi D_L}{b^2} \exp(-\beta u_F) \exp(\beta \frac{\mu b(1+\nu)}{3\pi(1-\nu)R_0}\Delta V) \tag{B.134}$$

Thus, with the indicated numerical values, one obtains:

$$\mathbf{v} \approx 3.10^5 \tag{B.135}$$

expressed in interatomic spacings per second.

Exercise 30. *Nucleation and growth of a dislocation loop*

1°) a) The loop can be considered as decomposed into sets of pairs of small segments opposed on a diameter of the loop, having opposite orientations, and the same Burgers vector. Equivalently, each pair of segments can be considered as having the same orientation and opposite Burgers vectors. At a large distance (much larger than the loop diameter ρ), the two segments of a pair generate opposite stress fields which cancel eachother (since the stress fields are proportional to the algebraic value of $\vec{b}$).

b) Aside from the two infinitesimal portions of circle perpendicular to $\vec{b}$ which are of the edge-type, and of the two infinitesimal portions parallel to $\vec{b}$ which are screw-type, the loop is constituted of mixed-type dislocations.

c) For a straight dislocation, the energy per unit length associated to the stress field generated by the dislocation is (Eq. (6.16)):

$$\frac{\mu b^2}{4\pi K} Ln(\frac{R}{r_0}) \tag{B.136}$$

in which R is the distance beyond which the field is vanishes, r_0 is the radius of the dislocation-core, and K is equal to 1 or to $(1-\nu)$ depending if the dislocation is of the screw-type or of the edge-type. Choose the x axis parallel to $\vec{b}$, and denote θ the polar angle of the current point of the circular loop. The infinitesimal portion $\rho d\theta$ of the dislocation has an edge-component of Burgers vector $b\cos\theta$ and a screw-component of Burgers vector $b\sin\theta$. One deduces, taking, as assumed, $R = \rho$:

$$E_e = \frac{\mu b^2}{4\pi}\rho Ln(\frac{\rho}{r_0}) \int_0^{2\pi} [\sin^2\theta + \frac{\cos^2\theta}{1-\nu}]d\theta = \frac{\mu b^2}{4\pi}\frac{2-\nu}{(1-\nu)}\pi\rho Ln(\frac{\rho}{r_0}) \tag{B.137}$$

d) A fluctuation $\delta\rho$ increases the elastic energy induced by the dislocation, by the amount δE_e:

$$\delta E_e = \frac{\mu b^2}{4\pi} \frac{2-\nu}{1-\nu} \pi [1 + Ln(\frac{\rho}{r_0})]\delta\rho \qquad (B.138)$$

The work done by σ corresponds to the gliding, by an amplitude b, of the surface between the circles of radii ρ and $\delta\rho$, equal to $2\pi\rho\delta\rho$. The stress exerts on this area the force $\sigma dS = 2\pi\sigma\rho\delta\rho$. Hence:

$$\delta W = 2\pi\sigma b\rho\delta\rho \qquad (B.139)$$

e) If $\delta U < 0$ for $\delta\rho > 0$, the loop will tend to enlarge under the action of the stress. It will tend to reduce if $\delta U < 0$ for $\delta\rho < 0$. The two situations correspond respectively to $\rho \gtrless \rho_C$, with :

$$\rho_C = \frac{\mu b}{8\pi\sigma} \frac{2-\nu}{(1-\nu)} [1 + \ln(\frac{\rho_C}{r_0})] \qquad (B.140)$$

For $r_0 = 3b$, $\sigma = 10^{-3}\mu$, and $\nu = 0, 5$, one can write:

$$\frac{\rho_C}{3b} = \frac{10^2}{8\pi} [1 + \ln(\frac{\rho_C}{3b}) \qquad (B.141)$$

Thus, with $X = (\rho_C/3b)$, one obtains the equation $(X/4) \approx (1 + LnX)$ which has two solutions (figure hereunder) one of which ($X < 1$) must be rejected since it corresponds to a loop smaller than the dislocation core. The other solution is $X \approx 15$ thus $\rho_C \approx 45b$.

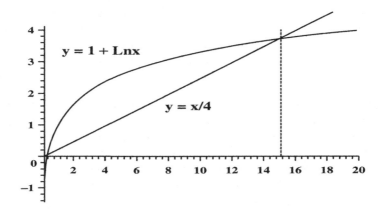

Figure B.19

f) In order to form a stable loop, the solid must have an available thermal energy equal to $U = E_e(\rho_C) - W(\rho_C)$ in which W is the work done during the formation of the loop, i.e. $(W = \pi\rho_C^2\sigma b)$. Then:

$$U = \mu b^3[0, 75.45. \ln(15) - \pi(45)^2 10^{-2}] \approx 27, 8\mu b^3 \approx 111\text{eV} \qquad (B.142)$$

Remark: Such a value corresponds to a probability $\exp(-U/k_BT) \approx 0$. A dislocation loop cannot therefore appear spontaneously in presence of a stress of the order of $10^{-2}\mu$.

Exercise 31. *Plasticity of a polycrystalline solid*

1°) The expression is (see chapter 7) $w = 2\sqrt{2bK/\mu} \approx 2b\sqrt{2C/\mu}$ $(C \approx K/b$ as indicated in the text). Hence:

$$\frac{w}{b} = 2\sqrt{\frac{2C}{\mu}} = 4 \qquad (B.143)$$

It results that:

$$\sigma_{PN} = 2\mu\exp(-\pi w/b) \approx 2\mu\exp(-12) \approx 1, 2.10^{-5}\mu \approx 0, 12\text{kg/mm2} \qquad (B.144)$$

2°) a) The required stress induces a curvature-radius R equal to half the length of the source (i.e. of the segment of dislocation pinned at its two ends). Then:

$$\sigma_{FR} = \frac{\alpha\mu b}{R} \qquad (B.145)$$

For an edge-dislocation, α is the coefficient of μb^2 in the expression of the energy density. Assuming that the range of of the stress field induced by the dislocation is equal to the average radius $(d/2)$ of a grain, the core radius being $r_0 = (w/2) = 2b$:

$$\alpha = \frac{Ln(d/4b)}{4\pi(1-\nu)} = 1.83 \qquad (B.146)$$

Thus:

$$\sigma_{FR} = \frac{1, 83.2, 5}{5.10^4}\mu = 0, 9.10^{-4}\mu \approx 0, 9\text{kg/mm}^2 \qquad (B.147)$$

b) The angular effect determines a ratio $\sigma_{FR} = \sigma_T.\cos\theta\sin\phi$. (cf. chapter 8). Hence:

$$\sigma_T^{(1)} = \frac{\sigma_{FR}}{\cos 45° \sin 45°} = 2\sigma_{FR} = 1.8\text{kg/mm}^2 \qquad (B.148)$$

3°) a) This results from the Frank-Read mechanism which generates dislocation loops all situated in the same glide plane.

b) The Peach-Koehler formula determines a force directed to the right and equal to $F_\sigma = \sigma b$.

c) The edge dislocations situated on the same side of the source all have the same sign. They therefore repel each other. As the repulsion force modulus diverges as $(1/r)$, each dislocation can block the progression of the following dislocations and set an appropriate distance between dislocations.

d) The n dislocations exert, on the dislocation source, repelling forces which oppose the "engine" stress σ which ensures the functioning of the source. The assumption made implies that the n dislocations are approximately at the distance $(d/2)$ from the source. The exerted force (directed to the left) is thus (Eq. (6.2) with $\theta = 0$):

$$F = n. \frac{\mu b^2}{2\pi(1-\nu)(d/2)} \tag{B.149}$$

The source will be blocked if $F \geq \sigma_{FR}.b$ (The effective stress applied to the source will then be smaller than the Frank-Read stress.) Thus:

$$n = \frac{2\pi(1-\nu)d\sigma_{FR}}{\mu b} \tag{B.150}$$

For $\sigma_{FR} = 0.9.10^{-4}\mu$:

$$n = \frac{2\pi.0.6.10^{-4}0.9.10^{-4}}{2.5.10^{-10}} \approx 140 \tag{B.151}$$

Actually, this number is an upper limit. Indeed as is shown further, 140 dislocations are distributed on a large segment, as compared to the size of a grain. Hence, several dislocations are close to the source, and exert an opposing force much larger than the one evaluated by assuming that all the dislocations are situated at the same distance $(d/2)$.

4°) a) The force exerted by the edge-dislocation x_i son the unit length of the dislocation x_j is determined by (Eq. (6.2)):

$$F_{ij} = \frac{\mu b^2}{2\pi(1-\nu)}.\frac{1}{x_j - x_i} \tag{B.152}$$

The equilibrium equation of the dislocation j can be deduced from F_{ij}:

$$F_j = \frac{\mu b^2}{2\pi(1-\nu)} \sum_{i \neq j} \frac{1}{x_j - x_i} = -\sigma b \tag{B.153}$$

b) Considering the form of W :

$$\sum_1^n \frac{\partial W}{\partial x_i} = 0 \tag{B.154}$$

One can deduce:

$$F_1 = -\frac{\partial W}{\partial x_1} = \sum_2^n \frac{\partial W}{\partial x_j} = -\sum_2^n F_j = (n-1)\sigma b \qquad (B.155)$$

Taking into account the external shear stress, the total force applied to the unit length of the head-dislocation is $n\sigma b$. An opposite force must be applied to the grain boundary to block this dislocation.

c) An argument similar to the one used to establish the Peach-Koehler formula can be used. Thus, a displacement δl of the n dislocations corresponds, for each dislocation, to an extension of the glide area by an amount $(1.\delta l)$ per unit length of the dislocation. The force exerted on this area is $\sigma \delta l$. It is displaced by b. Hence the total work done is $n\sigma b\delta l$. The effective force exerted on the rigid set of n dislocations is therefore equal to $n\sigma b$.

d) One must have $n\sigma b \geq F_G$, with $\sigma = 2\sigma_{FR}$ thus:

$$n = \frac{F_G}{2\sigma_{FR}.b} = \frac{20\sigma_{FR}.b}{2\sigma_{FR}.b} = 10 \qquad (B.156)$$

This number is much smaller than the number of dislocations needed to block the source, hence the source works continuously. The 10 first dislocations pile up in front of the grain boundary. The head dislocation then crosses the boundary. There are permanently a set of 10 dislocations in front of the grain boundary. Each additional dislocation generated by the source induces the crossing of the head dislocation to the adjacent grain.

5°) The equilibrium equation of the "tail" dislocation is (changing the signs in the equilibrium equation written above), with δ the distance between dislocations:

$$\frac{\mu b^2}{2\pi(1-\nu)} \sum_{i \neq n} \frac{1}{x_i - x_n} = \frac{\mu b^2}{2\pi\delta(1-\nu)} \sum_1^{n-1} \frac{1}{i} = \frac{2,8\mu b^2}{2\pi\delta(1-\nu)} = 2\sigma_{FR}.b \quad (B.157)$$

The length of the set of dislocations is then (with σ expressed in N/m^2):

$$9\delta = \frac{9.2,8\mu b}{4\pi(1-\nu)\sigma_{FR}} \approx \frac{85}{\sigma_{FR}} \text{meter} \qquad (B.158)$$

For $\sigma_{FR} = 0,9 \text{kg/mm}^2 \approx 9.10^6 \text{N/m}^2$, the length obtained is ≈ 10 microns.

6°) The maximum size of the set of dislocations which can pile up in front of the grain boundary is $(d/2)$. Their maximum number is then $(d/2\delta)$. Thus:

$$\frac{d}{2} \approx n_{max}\delta \approx n_{max}.\frac{\mu b}{2\pi(1-\nu)\sigma} \sum_1^{n-1} \frac{1}{i} = \frac{3n_{max}\mu b}{2\pi(1-\nu)\sigma} \qquad (B.159)$$

And:

$$n_{max} = \frac{2\pi(1-\nu)\sigma d}{6\mu b} \qquad (B.160)$$

On the other hand, the stress acting on the head dislocation is $n\sigma$. This stress must be at least equal to the stress F_G/b required to progress across the grain boundary. Therefore:

$$n_{\max}\sigma = \frac{2\pi(1-\nu)\sigma^2 d}{6\mu b} \geqslant \frac{F_G}{b} \tag{B.161}$$

Thus:

$$\sigma \geqslant \sqrt{\frac{6\mu.F_G}{2\pi(1-\nu)}}\cdot\sqrt{\frac{1}{d}} \tag{B.162}$$

This stress increases when the diameter of the grains decreases. For $d = 10$ microns, one has, with $F_G = 20\sigma_{FR}.b \approx 2.10^8.b$ (in N/m^2) :

$$\sigma \geqslant \sqrt{\frac{6\mu.20.\sigma_{FR}}{2\pi(1-\nu)}}\sqrt{\frac{b}{d}} = \sqrt{\frac{6.10^{11}.2.10^8}{2\pi.0,6}}\sqrt{\frac{2,5.10^{-10}}{10^{-5}}} \approx 2.10^7.\mathrm{N/m}^2 \approx 2\mathrm{kg/mm}^2 \tag{B.163}$$

7°) Consider first a single grain ($s \approx 10^{-4}\mathrm{cm}^2$) which contains, in average, a single mobile dislocation. By evacuating this dislocation, a step of amplitude b will occur, which produces an increment of shear deformation equal to $\epsilon = (b/d) \approx 2.5.10^{-6}$ smaller than the mentioned limit of observability. In order to be able to observe a plastic deformation, a stress σ_{FR} must be applied to start the functioning of the source and generate new dislocations. If a polycrystalline sample of 1cm^2 is considered, containing 10^4 mobile dislocations, the induced deformation is $\epsilon = 10^4 b/h$ (with $h = 1\mathrm{cm} = 10^2 d$). One obtains $\epsilon = 2.5.10^{-4}$, thus a deformation larger to the limit of observability. The yield stress will then be σ_{PN}. This result is ambiguous since it depends of the size of the sample.

It seems more reasonable to consider the polycrystalline sample globally, as the shear stress is applied, in practice, to the entire sample. This implies, however that the assumption of a "free-motion" of the dislocations between grains be true. As the grain boundaries are obstacles to this free-motion. The required stress will therefore be even larger than σ_{FR} if the grains are small enough.

Exercise 32. *Polygonization of a rod*

As shown in exercise 13 the interaction between edge-dislocations having the same Burgers vector and parallel glide planes, will have the effect of displacing them is such a manner as to position them in stable positions, one below the other. On the other hand if dislocations share the same glide plane, they will repell each other. The spatial distribution of dislocations will form "columns" of dislocations regularly spaced between A and B. Note that each column induces a disorientation between the portion of crystal situated at its left and that situated at its right (cf. exercise 8). The angle between the two disoriented parts of the crystal is proportional to the number n of dislocations in a column ($\theta = nb/BC$). The rod thus takes the shape of a polygon whose consecutive edges make an angle θ.

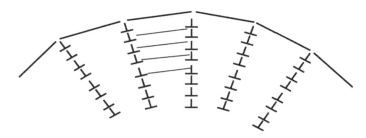

Figure B.20

Index